Defense Industrial Base Protection and Homeland Security

ALSO IN THE CRITICAL INFRASTRUCTURE AND HOMELAND SECURITY SERIES

Chemical Infrastructure Protection and Homeland Security
Nuclear Infrastructure Protection and Homeland Security
Water Infrastructure Protection and Homeland Security
Energy Infrastructure Protection and Homeland Security
Food Supply Protection and Homeland Security
Dam Sector Protection and Homeland Security
Transportation Protection and Homeland Security
Government Facilities Protection and Homeland Security

Defense Industrial Base Protection and Homeland Security

FRANK R. SPELLMAN

Bernan Press
Lanham • Boulder • New York • London

Published by Bernan Press
An imprint of The Rowman & Littlefield Publishing Group, Inc.
4501 Forbes Boulevard, Suite 200, Lanham, Maryland 20706
www.rowman.com
800-462-6420

Unit A, Whitacre Mews, 26-34 Stannary Street, London SE11 4AB

ISBN-13: 978-1-59888-994-9
eISBN-13: 978-1-59888-995-6

∞™ The paper used in this publication meets the minimum requirements of American National Standard for Information Sciences—Permanence of Paper for Printed Library Materials, ANSI/NISO Z39.48-1992.

Printed in the United States of America

For Kathern Welsh

Contents

Preface

The ninth of a new, well-received, and highly acclaimed series on critical infrastructure and homeland security, *Defense Industrial Base Protection and Homeland Security* is an eye-opening account and an important reference source for a complex sector. This book was designed and written to serve and advise U.S. project designers, engineers, law enforcement and security specialists, managers, building and grounds superintendents and/or supervisors, and responsible managers in charge of protecting the multifaceted nature of critical infrastructure in the United States. When I mentioned to a colleague of mine that I was in the process of writing the ninth volume of this series and I described the topic to be covered, she was surprised that I would embark on such a mind-numbing and difficult task: "How can you—or anyone—write about and describe the U.S. defense industrial base when it is so deep, tall, wide, all-encompassing, and almost indefinable in terms of total reach, scope, extent, and so on?" I remember scratching my head and trying to answer her in an impressive way (not easy) and finally coming up with: "If it exists, it can be written about and described." Then she scratched her head and smiled while shaking her head.

Anyway, currently, the defense industrial base critical infrastructure sector includes research and development, as well as design, production, delivery, and maintenance of military weapons systems, subsystems, and components or parts, to meet U.S. military requirements. Well, this part of the description is plain to see and easy to understand. Although it can deal with complex technology such as rocket science, it does not take a rocket scientist to comprehend the complexity of this sector. But hold on, it gets even more complex; there is more, much more. Consider that a U.S. sailor at sea or a marine or soldier or airperson can't (or is not able to) activate, target, and launch a weapons system unless he or she is trained properly to do so; and unless she or he is healthy and properly fed; and unless she or he is relatively comfortable in appropriate clothing and furnishings and environmental conditions (e.g., an HVAC-supplied environment). The point is these are just a few of any number of items and services provided to the defense department by the industrial defense base providers. Moreover, many defense industry base sector companies and their subcontractors provide incidental materials and services to the Department of Defense, including government-owner/contractor-operated and government-owned/contract-operated facilities. Note that these services do not include the commercial infrastructure of providers of services such as power, communications, transportation, or utilities that the Department of Defense uses to meet military operational requirements. These commercial infrastructure assets are addressed by other Critical Infrastructure Sector-Specific Agencies—these are discussed in detail in the other volumes of this series.

This book is organized to simplify and present in a logical and sequential manner a discussion of not only the elements comprising the defense industrial base sector but also many of the security measures employed to protect the various entities involved.

Let's face today's reality: those who want quick answers to complicated questions—to help employers and employees handle security threats—must be prepared to meet and deal with the threat of terrorism on a 24/7 basis. It is important to point out that this book does not discuss and focus on security concerns related to natural disasters; on the contrary, the focus here is on the security aspects in the design of systems that need to be able to deal robustly with possible sources of disruption, specifically from malicious acts. Moreover, the focus includes the added dimension of preventing misuse and malicious behavior. In the post–September 11, 2001, world the possibility of government facilities infrastructure terrorism—the malicious use of weapons and cyberintrusion to cause devastating damage to defense industrial base infrastructure and its associated subsectors along with—literally—its cascading effects—is very real. Thus, the need is clear and real and so is the format and guidelines presented in this text to improve protection and resilience of defense industrial base infrastructure.

This book describes the sector- and subsector-wide process required to identify and prioritize assets, assess risk in the sector, implement protective programs and resilience strategies, and measure their efficacy. This book (and all the volumes of the critical infrastructure sector series) was written as a result of 9/11 to address these concerns. It is important to point out that our defense industrial base infrastructure (as is the case with the other thirteen critical infrastructures) cannot be made absolutely immune to all possible intrusions/attacks; thus, it takes a concerted, well-thought-out effort to incorporate security upgrades in the retrofitting of existing systems and careful security planning for all new facility infrastructure components. These upgrades or design features need to address issues of monitoring, response, critical infrastructure redundancy, and recovery to minimize risk to the facility infrastructure. However, based on personal experience none of these approaches is or can be effective unless defense industrial base sector staff members at all levels of the chain of command are cognizant of the threats.

Defense Industrial Base Protection and Homeland Security presents common-sense methodologies in a straightforward manner. This text is accessible to those who have no experience with or knowledge of the defense industrial base sector. If you work through the text systematically, you will gain an understanding of the challenge of domestic preparedness—that is, an immediate need for a heightened state of awareness of the present threat facing the defense industrial base sector members as potential terrorist targets. Moreover, you will gain knowledge of security principles and measures that can be implemented—not only adding a critical component to your professional knowledge but also giving you the tools needed to combat terrorism in the homeland—our homeland, both by outsiders and insiders.

One final word to readers: This book is written in the conversational and engaging and reader-friendly style that is the author's trademark. Why? Well, when demonstrating how one—or anyone—can write about the U.S. defense industrial base when it is so deep, tall, wide, all-encompassing, and almost indefinable in terms of total reach, scope, extent, and so on, I never apologize for attempting to communicate.

Acronyms and Abbreviations

APM	Asset Prioritization Model
BIS	Bureau of Industry and Security
CII	Critical Infrastructure Information
CIKR	Critical Infrastructure and Key Resources
CIP	Critical Infrastructure Protection
CS/IA	Cybersecurity/Information Assurance
DCIP	Defense Critical Infrastructure Program
DHS	Department of Homeland Security
DIB	Defense Industrial Base
DOC	Department of Commerce
DoD	Department of Defense
DoDD	Department of Defense Directive
DoDI	Department of Defense Instruction
DOE	Department of Energy
DOS	Department of State
DPAS	Defense Priorities and Allocations System
EPA	Environmental Protection Agency
FEMA	Federal Emergency Management Agency
FOIA	Freedom of Information Act
HSPD	Homeland Security Presidential Directive
IA	Information Assurance
IP	Office of Infrastructure Protection

IT	Information Technology
NCIP	National Critical Infrastructure Protection
SSA	Sector-Specific Agency
SSP	Sector-Specific Plan
TSA	Transportation Security Administration

Prologue

The Cookie-Cutter Caper

Note: The following fictitious account is included here in order to set the stage for what follows in the text and to continue the sordid saga of the Williams family. The Williams are homegrown radicalized terrorists (smart and brutal and not Muslim) who have done whatever they could do to bring about the death and destruction of innocents in the United States, with particular attention being paid to attacking, impacting, and/or destroying sectors of critical infrastructure.

*****

I just hate Christmas! Don't you?

All those phony trappings . . . wow . . . what a waste. Anyway, anytime I think about holly so green, snow-covered rooftops, all those bright-colored lights casting their glare here and there and everywhere I just want to puke. Don't you?

And then there is that disgusting jolly old fat man dressed up in red and yelling out that sickening refrain: Ho, Ho, Ho! To me that is Oh So Low.

Then there are all those trees . . . fake trees and real ones . . . what a total waste. Fake trees are as fake as the commercial event they represent . . . and the real trees . . . can you believe the fools out there that actually have the audacity to go out to the woods and kill a tree and move the poor destroyed thing into their homes. Then they throw all kinds of ridiculous ornaments and lights on those sagging, drying, dying boughs. The only good news is when the lights on those dying boughs cause a fire and burn down their house and sometimes the occupants . . . now that is a Christmas event worth celebrating.

Oh, do not forget the children . . . those poor brainwashed, innocent children . . . What society does to them . . . to children . . . stuffing them with all that Christmas glitz and those terrible presents. Now, I will admit, some families do give presents that make sense. In my case, (yes, I admit we celebrated that horrible event . . . but only to a point, you understand) for instance, when I was five my father gave me a Model 84 Winchester Classic 30-30 rifle and box of bullets. I kept that treasured killing machine with me until my latest arrest . . . that is, until the feds confiscated it. In addition, one of my oldest brothers gave me a Bowie knife . . . had that one for years . . . so sharp it would almost cut anything . . . and it did . . . many times.

Well, let's get back to the baloney. Actually, I do have a weakness for Christmas cookies . . . and it is because of those cookies that I am presently locked up in this cell in King County Lock Up here in Seattle . . . my hometown . . . where all the bleeding heart, liberal dumb bunnies live. I look around at the bars on the cell door and the three windowless walls painted sickening pea green. My gaze settles on the initials and short, pithy statements carved into these walls (I can't repeat those statements here, folks, even I have some modesty). Most of the initials I am familiar with . . . many belong to my blood: I am part of that famous . . . excuse, me, infamous . . . family, the Williams. I am BM Williams V—and proud of it. I know you have heard of us and our famous and patriotic exploits, which unfortunately end with so many of us inhabiting this isolation cell sooner or later before we have a chance to kill as many of you no-loads as possible. Many of us are in jail, but be advised, we are a large family and are just getting started in our cleansing of America.

Okay, now that all you poor souls out there in la-la land have been introduced to me you are foaming at the bit to find out what it is that I have done that caused me to end up in this pea-green dungeon.

It began as a brilliant plan—or as that lousy DA called it, The Cookie-Cutter Caper. You see, two things me and my blood share is our hatred for anything federal or state or local that has anything to do with enforcing all these stupid laws the commies put into effect . . . and our innate genius. Einstein, Newton, Aristotle, and da Vinci—they were pretty smart, for sure, but the Williams are just a whole lot smarter. We just have bad luck . . . lots of bad luck and incompetent people.

I wanted to follow the old family tradition and destroy as much property and kill off as many despicable humans as possible. A few years ago, I came up with a plan to destroy part of the defense industrial base sector. This was and is no easy target because this sector is so huge and widespread. I had to whittle down to a specific target . . . one that was doable, destroyable. I decided to attack a huge part of the aerospace branch of the defense industrial base. Now, be advised I can't mention any specific industry names because that would not go good for me in court with my appeal . . . etc. Anyway, my lawyer said: "Woe be it to anyone stupid enough to mention names . . . targets . . . before her day in court." She clearly emphasized this point over and over again.

Anyway, I knew that I could not attack the entire defense industrial base . . . even with a large gang of fellow patriots helping. And, based on family experience, I have found that if you rob anything, destroy anything, or are crooked and bad to the bone marrow like me, work alone . . . always work alone . . . always. You understand when you hook up with others, your association becomes a chain and the others are all weak links. Many of my family members are incarcerated, deceased, or in one of them American nut wards because they did not work alone. Lone wolves forever! That is my mantra.

Anyway, I came up with a brilliant plan. Based on Christmas, actually . . . I know that Americans are gullible and mostly stupid. Americans lose their self-control and what little smarts they have during Christmas . . . they get emotional and downright teary and so gullible. Do you follow me? Good. Well, being a genius and innately smart in all areas of science, I had invented a drug like LSD that when ingested even in a pinhead-like quantity, it turns the average dumb American into either a Jack the Ripper or Jackie the Ripper type. It allows personality disorder on super steroids, my friends. So, I decided I would make Christmas cookies. Yes, plain sugar cookies cut in the shapes of snowflakes, angels, stars, reindeer, and Christmas trees. The cookie cutter that did me in was the Christmas tree . . .

I have to admit that I made mistakes . . . all geniuses do . . . we make mistakes. My first mistake was with the Christmas tree cookie cutter. It was made of cheap metal and manufactured in one of those foreign countries . . . you know the one . . . the one that poisoned our dog food, toothpaste, and drywall . . . and who knows what else. Anyway, I must have rolled out acres of cookie dough on several tables. So, it took me three days to cut out and then bake thousands of cookies of all shapes . . . etc. I made my first mistake while cutting the cookie dough. I used the Christmas tree cutter and because the metal cutter was so cheaply made one of its edges broke off. I was so infuriated with the faulty cookie cutter that I threw it across my garage and quickly forgot it.

My next mistake had to do with personal vanity and my huge ego. Being a lady with flaming red hair, green eyes, a sprinkling of freckles here and there, and a body that would make even Marilyn Monroe envious, I figured my good looks would keep the idiots at the two aerospace factories I picked spellbound. So I did not disguise myself—okay, yes, big mistake.

I set up several tables at the first aerospace location (I never got to the second one) and placed several thousand poisoned cookies on holiday-themed table cloths. Unfortunately, the tables where I put the cookies were directly under one of their CCTV systems. Thus, my non-disguised self was clearly exposed and recorded by their cameras. Also, the first person to pick up a couple of the cookies was a security guard (actually a Homeland Security copper) who had her guard dog right beside her. Before she could take a bite of the cookie she dropped one to the ground and the dog devoured it. Almost instantly, the dog went into convulsions—and died a horrible, mouth-foaming death. Well, not all cops or security guards are stupid: the guard immediately was suspicious, and before I could escape she threw me to the ground and handcuffed me. And that is how I ended up here, in lockup.

After I had been woman-handled and thrown in jail, the suits still had no hard evidence that I had done anything wrong. But that quickly changed. The Homeland Security types and others confiscated all those cookies because they figured correctly that I had poisoned them. I felt sorry for the dog . . . dogs are always better than humans . . . At first the liberal media tore into the cops for my rough treatment, which was recorded plainly on a passerby's smartphone and given to the press. You know those liberals, they hate the police . . . despise them . . . so, at first I was okay. That is, until they did a lab test—and inspected my house and garage which I had cleaned up with no trace evidence . . . Or so I believed. But I had forgotten one item: that damaged Christmas tree cookie cutter that I had thrown across the garage and then forgot about. That cookie cutter had traces of the same poison the lab found in my cookies.

So, I was and am now a cooked goose, so to speak. The good news is I killed no one so my liberal lawyer thinks I will get off easy in this bleeding heart town and I will be free again. Free to bring about mayhem, death and destruction . . . and I will not be denied. You have been forewarned: You will hear from BM Williams V again.

1

Introduction

They merely act on impulse, seeking what they think they want, glory, power, wealth, the trappings of might.

—Raymond E. Feist, *Shards of a Broken Crown*

In America, there are plenty of hate groups that claim peace and brotherhood, but when their actions are responsible for death and destruction, they are identified for what they really are.

—Lindsey, *Vocabulary of Hate*

JUST TAKE WHAT WE NEED!

Over the millennia, one of the primary questions that every army and navy commander has asked himself and others is, how do we feed, clothe, and rearm our soldiers and sailors? In the days of the barbarians and Vikings the answer to this question was easy: invade a city or human gathering area and kill the occupants, except for those destined for slavery, and then confiscate food, clothing, and any weapons they could find. Problem solved. Even in relatively modern times, in the U.S. Civil War, for example, armies confiscated food, clothing, and weapons. For Sherman's famous March to the Sea in 1864, the question today might be: How did General Sherman feed and clothe his approximately 65,000 union troops while they raided from Atlanta to Savannah, Georgia, during his five-week march to the sea? The answer is simple: Sherman's army confiscated what they needed and wanted and destroyed all that remained. Today this "hard war" strategy is studied and admired throughout the globe. The fact is that in wartime, invasion and confiscation of the enemy's goods are tactics that are practiced to this very day. An army and navy fights on its stomach. Common sense tells us that an army and navy that are well fed, properly clothed, and adequately armed are not only significant fighting forces but generally healthy ones, too.

At the present time and state of affairs worldwide (with a few exceptions), another more pertinent question is how to keep a *peacetime* army, navy, and air force supplied with its daily needs in respect to food, clothing, housing, training, equipment, armament, materials, and medical support and services? No easy task, for sure. But experience has shown it is doable. Those providing the doing are what this book is all about; that is, protecting the doers from terrorist acts. The Defense Industrial Base (DIB) is the supplier of the military's needs. So, the next questions are, what is the DIB sector? And what is it composed of? DIB consists of more than 250,000 firms in 215 industries. The fact is our nation's defense and military strength relies primarily on the Department of Defense (DoD) and the private-sector defense industry that supports it. Without the important contributions of the private sector, DoD can't effectively execute its core defense missions, including mobilization and deployment of our nation's military forces abroad. Conversely, private industry and public at large rely on the federal government to provide for the common defense of our nation and protect our interests both domestically and abroad (DHS 2010).

If we are to be successful in the war on terrorism, we must have a military that has the ability to mount swift, calculated offensive and defensive operations. This capability can be achieved only if our military is well trained and properly equipped. The majority of the equipment, materials, services, and weaponry used by our military forces are provided by private industry. DoD, for decades, has worked to identify its own critical assets and systems. It has also begun to address its dependency on the defense industrial base, and is now taking the concerns of private industry into consideration in its critical infrastructure protection assessment efforts.

It should be noted that recent trends driven by sequestration legislation and budget constraints, in general, along with market competition, consolidations, globalizations, and attrition have reduced or eliminated redundant sources of products and services and therefore increased risk for DoD. Complex domestic and foreign corporate mergers and acquisitions and outsourcing have made it even more difficult for DoD to be assured that its prime contractors' subcontractors understand its security requirements and are prepared to support them in a national emergency.

In looking at the history of DIB, especially over the past thirty years, one thing is certain; DOD's dependency on the private sector has greatly increased. Because of outsourcing DoD has relied increasingly on contractors to perform many of the tasks that were once under the exclusive purview and control of the military. This is the case even with vital utilities that service many of the nation's important military installations. Why? Because many of the utilities are privatized or in the process of being privatized, reliance on single-source providers and a very limited number of private-sector suppliers has become the norm in provision of its most essential needs. Unlike other federal government agencies, DoD requires strict adherence to military product specification and unique requirements for services. DoD's unique requirements and specifications may make supplying such needs difficult and restricted to very few suppliers in the world who have the capability of satisfying these rigorous requirements and needs. Because many of these specialty suppliers have single manufacturing and distribution points, they justify additional and continuous security review and assessment.

Another bugaboo for DoD is related to its current process by which it contracts with the private sector to provide critical services and suppliers. Because the DoD procurement process most often is based on cost and efficiency, this approach may not always take into account the vendor's critical infrastructure protection-practices (e.g., site security, workplace hiring, supplier base) and its ability to supply products and services and provide surge response during an emergency or exigent (demanding) circumstances (DHS 2003). This potential problem is compounded by growing concerns within the private sector regarding the potential for additional costs and risks resulting from federal mandates that require private industry to implement expensive enhanced, and often cumbersome, protection measures.

The Department of Homeland Security understands the importance of DIB and the associated costs imposed on them to comply with federal security requirements. Because of the importance of DHS and DIB infrastructures they have been integrated on many levels. DoD, in concert with DHS, continues working with the private sector to identify critical installations and, subsequently, to delineate specific protection requirements (DHS 2003). In addition, DoD and DHS collaborate with key defense industrial base organizations in integrating and building upon their individual existing protection plans.

Additional defense industrial base protection initiatives have been put into place including efforts to build critical infrastructure protection requirements into contract process and procedures. DoD collaborates with the defense industry to review contract processes and procedures to determine how to include provisions that address critical infrastructure protection needs. Contracts specifically address national emergency situation requirements, such as contractor response times, supply and labor availability, and direct logistic support. When appropriate, contracts also include language regarding program manager accountability for the protection of supporting infrastructures. Sensitive contractual documents receive greater scrutiny and revision prior to public posting. Additionally, DoD gives specific scrutiny to its potential dependency on foreign commercial operators and suppliers. Another protection initiative involves the incorporation of security concerns into production and distribution process and procedures. DoD and industry explore ways to eliminate key production and distribution bottlenecks. Finally, an effective means of sharing security-related information between defense organizations and private-sector service providers has been developed. DoD works with DHS and the intelligence and law enforcement communities to establish the necessary polices and mechanisms to facilitate a productive exchange of security-related information with the defense industry.

TERRORIST ATTACKS AGAINST AMERICANS

It was not that far in America's distant past that the dangers we dealt with were primarily the ramifications of natural disasters. Although never welcomed, these natural occurrences were expected, planned for, guarded against, and recovered from. Of course, there have always been human-caused disasters that we also have had to contend with. The sinking of the *Titanic* in 1912 prompted new safety regulations for passenger ships; the fire at New York City's Triangle

Shirtwaist Factory in 1911 that took the lives of 146 laborers—mainly young women—was the catalyst for the development of workplace safety and fire regulations. Still, the ultimate wake-up call for Americans came in the form of the events that we now summarize as "9/11." On September 11, 2001, terrorists struck at the heart of America—on American soil in a way that was unforgettable. Airplanes filled with people and fuel were turned into guided missiles of death and destruction.

9/11 was a wake-up call that alerted us to a new type of hateful venom—coldly delivered by groups or individuals in order to kill massive numbers of people, cause substantial property damage, and affect economic stability. Most importantly, for the public and emergency responders, the events of 9/11 brought awareness that just about anything unthinkable is possible. The bottom line: security of the homeland is vital to all of us.

That great philosopher (Bob Dylan, of course) noted that "The Times They Are a-Changin,'" and recent terrorist events have made this well-known phrase certainly the case. For example, with the manifestation of al Qaeda, the Islamic State of Iraq and Syria (ISIS), also known as the Islamic State of Iraq and the Levant (ISIL) and Daesh, and other terrorist groups—including domestic fanatics—there are many out there that who would kill us all. It is important to point out with the advent of insider/domestic threats, the message of this book is the same as in the other books in this series where the focus has been shifted to the lone wolf terrorist. The fact is that there are homegrown terrorists who live among us—plotting, quivering in the glory of mass murder and destruction, and envisioning the attended glory of making the headlines as martyrs in the sickest sense possible. These are people who live among us, who shop in our stores, who use the benefits of our free society to grow their hate and to enhance their total disgust for all those things that we value in life.

Before 9/11, terrorism was not previously seen as a significant threat to the United States. However, while the United States has not experienced continuous 9/11-type terror events in the past, the current climate suggests that the United States is at significant risk of further terror activity in the future. Forecasting the future is impossible, but we can look at the past and learn from it. Table 1.1 summarizes past acts of sabotage/terrorism carried out in the United States, or against U.S. citizens overseas.

Table 1.1. Selected Terror Attacks since 1993 (within the United States or against Americans abroad)[1]

Date/Location	*Summary of Attacks*
Feb. 26, 1993 New York City	World Trade Center truck bomb in garage. Six deaths and 1,042 injuries. Intended structural collapse did not occur. Several members of Middle Eastern extremist organizations convicted of roles in the bombing.
April 19, 1995 Oklahoma City	Truck bomb exploded outside Alfred P. Murrah Federal Building, collapsing walls and floors. 168 people were killed. More than 300 other buildings sustained damage. Timothy McVeigh and Terry Nichols later convicted in the anti-government plot to avenge the Branch Davidian standoff in Waco, TX, exactly 2 years earlier.
Nov. 13, 1995 Riyadh, Saudi Arabia	Car bomb exploded at U.S. military training facility, killing 6, including 5 U.S. military servicemen.
Oct. 9, 1995 Hyder, Arizona	Amtrak Sunset Limited passenger train traveling from New Orleans to Los Angeles carrying 248 people derailed, killing one and injuring 78. Investigations indicated that tracks had been deliberately tampered with by one or more who left anti-government notes at the site.
June 25, 1996 Dhahran, Saudi Arabia	Truck bomb exploded outside Khobar Towers housing complex, killing 20, including 19 American servicemen, and injuring hundreds of others. Alleged members of Hezbollah were indicted on charges relating to the attack in June 2001.
July 27, 1996 Atlanta, Georgia	Bomb exploded at Centennial Olympic Park during the 1996 Summer Olympics killing one and injuring 111. Domestic terrorist convicted.
Aug. 7, 1998 Nairobi, Kenya and Dar es Salaam, Tanzania	Truck bombs exploded almost simultaneously near two U.S. embassies, killing 224 (213 in Kenya and 11 in Tanzania) and injuring about 4,500. Attributed to al Qaeda. Indictments were brought against 22 men—including Osama bin Laden—but only four men were convicted of the killings in May 2001 and later sentenced to life in prison.
Sept. 11, 2001 New York City, Washington, DC, and Pennsylvania	Al Qaeda terrorists hijacked four commercial jet airlines. Two aircraft used against World Trade Center Towers in NYC, third against Pentagon. Fourth aircraft brought down by passenger intervention. Approximately 3,000 killed.
Sept 18–Oct 9, 2001 U.S.	U.S. postal system used as delivery vehicle for anthrax spores contained in letters. News organizations and government representatives were targeted. Five people died and 17 suffered effects.
May 12, 2003 Riyadh, Saudi Arabia	Suicide bombers killed 35, including 9 Americans, at housing compounds for Westerners. Al Qaeda suspected.
Dec. 6, 2004 Jeddah, Saudi Arabia	Terrorists stormed the U.S. consulate, killing 9 consulate employees.

Date/Location	Summary of Attacks
Nov. 9, 2005 Amman, Jordan	Coordinated bombings hit three hotels—Radisson SAS Hotel, Grand Hyatt Hotel, and Days Inn—in Amman, Jordan, killing more than 60 people and injuring 115. Al Qaeda claimed responsibility.
Sept. 13, 2006 Damascus, Syria	Syrian terrorists attacked the U.S. embassy, but Syrian security forces intervened to stop it.
Jan. 12, 2007 Athens, Greece	The U.S. embassy was fired on by an anti-tank missile causing damage but no injuries.
Dec 11, 2007 Algiers, Algeria	41 people were killed, including 17 United Nations staff members, when Al Qaeda terrorists detonated two car bombs near the Constitutional Court and United Nations offices.
July 13, 2008 Quam, Afghanistan	Nine U.S. soldiers died when Taliban militants boldly attacked an American base bordering Pakistan.
Sept. 16, 2008 Sana, Yemen	A car bomb and rocket struck the U.S. embassy in Yemen as staff arrived to work, killing 12 people. At least 25 suspected al Qaeda militants were arrested for the attack.
Nov. 26–29, 2008 Mumbai, India	Pakistani terrorists targeting sites popular with Americans and other foreign tourists attacked two five-star hotels, a hospital, a train station, and a cinema. More than 300 people were wounded and 161 people died, including at least 5 Americans.
June 1, 2009 Little Rock, Arkansas	Abdulhakim Muhammed, a convert to Islam from Memphis, Tennessee, shot two soldiers outside a military recruiting center, killing one. He pled allegiance to al Qaeda.
Dec. 25, 2009	A Nigerian man on a flight from Amsterdam to Detroit ignited a hidden chemical explosive but the fire was suppressed by two other passengers. The would-be bomber claimed allegiance to al Qaeda.
Dec. 31, 2009 Afghanistan	A suicide bomber killed eight CIA employees at a military base in Khost Province, Afghanistan. The Taliban claimed responsibility.
May 1, 2010 New York City	A car bomb was discovered in Times Square that failed to detonate. Faisal Shahzad pled guilty to 10 terrorism and weapons charges.
May 10, 2010 Jacksonville, Florida	A pipe bomb exploded while approximately 60 Muslims were praying in the Islamic Center of Northeast Florida. The attack caused no injuries.
Sept. 11, 2012 Benghazi, Libya	Militants armed with antiaircraft weapons and rocket-propelled grenades fired upon the American consulate, killing U.S. ambassador to Libya Christopher Stevens and three other embassy officials.
April 15, 2013 Boston, Massachusetts.	Pressure cooker bombs exploded near the finish line of the Boston Marathon. Three people were killed and more than 260 people were injured. Three days later, the two suspects—brothers from Chechnya—went on a shooting/robbing spree in an attempt to escape arrest, killing two police officers.

(*continued*)

Table 1.1. ***Continued***

Date/Location	*Summary of Attacks*
August 19–Sept 2, 2014	ISIS murdered two American journalists—James Foley and Steven Sotloff—by decapitation after holding them hostage.
Dec. 2, 2015 San Bernardino, California	Fourteen people were killed and more than 20 wounded when husband and wife Syed Rizwan Farook and Tashleen Malik opened fire at an employee holiday partly at the Inland Regional Center, San Bernardino, California. The suspects, killed later in a shootout, claimed to be inspired by ISIS.
June 12, 2016 Orlando, Florida	A gunman who claimed allegiance to ISIS in a 911 call opened fire at an Orlando nightclub, leaving 49 people dead and injuring 58. He was targeting the LGBT community. The gunman died at the scene.
Oct. 31, 2017 New York City	A "self-radicalized" terrorist supporting ISIS ran a van down the Hudson River Park bike path, killing 8 people and injuring 11.

1. Adapted from *Terrorist Attacks in the U.S. or Against Americans.* @http://www.infoplease.com/ipa/A0001454.html.

CRITICAL INFRASTRUCTURE—TERMS AND CONCEPTS

Every branch of science, every profession, every engineering process, and everything else in life has its own key concepts and language for communication. Critical infrastructure is no different. To work even at the edge of understanding critical infrastructure and how to protect it and guard against terrorism, you must acquire a fundamental knowledge of the key concepts and vocabulary related to the concepts and processes involved.

In this section, we introduce the key concepts and define the terms used by critical infrastructure practitioners in applying their skills to make their technological endeavors bear fruit. Although there is an extensive glossary at the end of the book, I'd like to present concepts and terms early in the text so readers can become familiar with them. The practicing security engineer, manager, practitioner, or student of protecting the country's critical infrastructure should understand these concepts and terms.

What Is Terrorism?

A few terms often are not only poorly defined but also defined from different and conflicting points of view. The definition of **terrorism** is an example.

After reviewing several dictionaries (too many to list here), we have found this fairly standard definition of terrorism:

> The unlawful use or threatened use of force or violence by a person or an organized group against people or property with the intention of intimidating or coercing societies or governments, often for ideological or political reasons.

America's *National Strategy for Homeland Security* defines terrorism as follows:

> Any premeditated, unlawful act dangerous to human life or public welfare that is intended to intimidate or coerce civilian populations or governments (White House 2006).

The U.S. State Department defines terrorism thus:

> Premeditated, politically motivated violence perpetrated against noncombatant targets by subnational groups or clandestine agents (U.S. Congress 2005).

The FBI definition of terrorism is:

> The unlawful use of force or violence against persons or property to intimidate or coerce a government, the civilian population, or any segment thereof, in furtherance of political or social objectives (FBI 2006).

Note that the FBI divides terrorism into two categories: domestic (homegrown), involving groups operating in and targeting the United States without foreign direction; and international, involving groups that operate across international borders and/or have foreign connections.

Well, at this point the obvious question is: Do you now know what terrorism is? That is, can you definitively define it? If you can't define it, you are not alone—not even the U.S. government can definitively define it. Maybe we need to look at other sources—views from the real experts on terrorism.

Osama bin Ladin's View on Terrorism

> Wherever we look, we find the United States as the leader of terrorism and crime in the world. The United States does not consider it a terrorist act to throw atomic bombs at nations thousands of miles away [Japan during World War II], when those bombs would hit more than just military targets. Those bombs rather were thrown at entire nations, including women, children, and elderly people . . . (Bergen 2002, 21–22)

Another View

This view is from court testimony on terrorism from Ramzi Ahmed Yousef, who helped organize the first terrorist attack on the World Trade Center in 1993:

> You keep talking also about collective punishment and killing innocent people to force governments to change their policies; you call this terrorism when someone would kill innocent people or civilians in order to force the government to change its policies. Well, when you were the first one who invented this . . .
>
> You were the first one who killed innocent people, and you are the first one who introduced this type of terrorism to the history of mankind when you dropped an atomic bomb which killed tens of thousands of women and children in Japan and when you killed over a hundred thousand people, most of them civilians, in Tokyo with fire bombings.
>
> You killed them by burning them to death. And you killed civilians in Vietnam with chemicals as with the so-called Orange Agent. You killed civilians and innocent people, not soldiers, innocent people every single war you went. You went to wars more than any other country in this century, and then you have the nerve to talk about killing innocent people.
>
> And now you have invented new ways to kill innocent people. You have so-called economic embargo which kills nobody other than children and elderly people, and which other than Iraq you have been placing the economic embargo on Cuba and other countries for over 35 years . . .
>
> The government in its summations and opening said that I was a terrorist. Yes, I am a terrorist and I am proud of it. And I support terrorism so long as it was against the United States government and Israel, because you are more than terrorists; you are the one who invented terrorism and using it every day. You are butchers, liars and hypocrites. (excerpts from court 1998)

And finally, here is an old cliché on a terrorist: One man's terrorist is another man's freedom fighter.

Again, from the preceding points of view, it can be seen that defining terrorism or the terrorist is not straightforward and never easy. Even the standard dictionary definition leaves us with the vagaries and ambiguities of other words typically associated with terrorism, such as in the definitions of **unlawful** and **public welfare** (Sauter and Carafano 2005).

Raphael Perl (2004) in a Congressional Research Service (CRS) report points out that one definition widely used in U.S. government circles, and incorporated into law, defines **international terrorist** as terrorism involving the citizens or property of more than one country.

Terrorism is broadly defined as politically motivated violence perpetrated against noncombatant targets by subnational groups or clandestine agents. For example, kidnapping of U.S. birdwatchers or bombing of U.S.-owned oil pipelines by leftist guerrillas in Columbia would qualify as international terrorism. In 22 U.S.C. 2656f, a **terrorist group** is defined as a group which practices terrorism or has significant subgroups that practice terrorism. Perl points out that one of the shortfalls of this traditional definition is its focus on groups and its exclusion of individual ("lone wolf") terrorist activity, which has recently risen in frequency and visibility.

At this point, readers may wonder, "Why should we care what the definition of terrorist or terrorism is?" Definitions are important because in order to prepare for the terrorism contingency, domestic or international, we must have some feel, as with any other problem, for what it is we are dealing with. We are fighting a war of ideas. We must attempt to understand both sides of the argument, even though the terrorist's side makes no sense to an American or other freedom-loving occupant of the globe.

The *Washington Times* (2009) reported that in an interview with German magazine *Der Spiegel*, then Secretary of Homeland Security Napolitano said that she rejected the word "terrorism," and has instead renamed terrorist acts using the euphemism "man-caused disasters," because "it demonstrates that we want to move away from the politics of fear, toward a policy of being prepared for all risks that can occur." This is analogous to stating that we should refer to serial killers, serial rapists, and serial arsonists as man-caused afflictions. After all, we do not want to create fear about serial offenders. Get real! The author of this text suggests that former Secretary Napolitano read the remarks we mentioned earlier by Ramzi Ahmed Yousef: "Yes, I am a terrorist and I am proud of it. And I support terrorism as long as it was against the United States government and Israel."

Finally, while it is difficult to pinpoint an exact definition of terrorism (but not difficult to rename it and call it something else), we certainly have little difficulty in identifying it when we see it, when we feel it, when we suffer from it. By any other name terrorism is best summed up as an absolute feeling of Terror—nothing judgmental about that—just Terror with a capital *T*.

For my purposes, probably the best definition of terrorism is given in U.S. Department of Homeland Security's *Defense Industrial Base Sector-Specific Plan* (2010):

> Terrorism is the calculated use of unlawful violence or threat of unlawful violence to inculcate [instill] fear, intended to coerce or intimidate governments or societies in the pursuit of goals that are generally political, religious, or ideological.

Other Important Definitions

In addition to the definition of terrorism I include a few of the important definitions related to this discussion of critical infrastructure in the following (DHS 2010). (See also the appendix for more terms.)

All hazards—A grouping classification encompassing all conditions, environmental or manmade, that have the potential to cause injury, illness, or death; damage to or loss of equipment, infrastructure services, or property; or alternatively causing functional degradation to social, economic, or environmental aspects.

Asset (Infrastructure)—A distinguishable network entity that provides a service or capability. Assets are people, physical entities, or information located either within or outside the United States and owned or operated by domestic, foreign, public, or private-sector organizations. DHS states an asset is a person, structure, facility, information, material, or process that has value.

Consequence—The effect of an event, incident, or occurrence. For the purposes of this book, consequences are dived into four main consequences: public health and safety, economic, psychological, and governance impacts.

Control systems—Computer-based systems used within many infrastructures and industries to monitor and control sensitive processes and physical functions. These systems typically collect measurements and operational data from the field, process and display the information, and relay control commands to local or remote equipment or human-machine interfaces (operators). Examples of types of control systems include Supervisory Control and Data Acquisition systems, process control systems, and distributed control systems.

Critical infrastructure—Systems and assets, whether physical or virtual, so vital that the incapacity of destruction of them may have a debilitating impact on the security, economy, public health or safety, environment, or any combination of these factors, across any federal, state, regional, territorial, or local jurisdiction.

Critical infrastructure information (CII)—Information that is not customarily in the public domain and is related to the security of critical infrastructure or protected systems, CII consists of records and information concerning any of the following:

- Actual, potential, or threatened interference with, attack on, compromise of, or incapacitation of critical infrastructure or protected systems by either physical or computed-based attack or other similar conduct (including the misuse of or unauthorized access to all types of communications and data transmission systems) that violates federal, state, or local law; harms the interstate commerce of the United States; or threatens public health or safety.
- The ability of any critical infrastructure or protected system to resist such interference, compromise, or incapacitation, including any planned or past assessment, projection, or estimate of the vulnerability of critical infrastructure of a protected system, including security testing, risk evaluation thereto, risk management planning, or risk audit.
- Any planned or past operation problem or solution regarding critical infrastructure or protected systems, including repair, recovery, insurance, or continuity, to the extent that it is related to such interference, compromise, or incapacitation.

Cybersecurity—The prevention of damage to, unauthorized use of, or exploitation of, and, if needed, the restoration of electronic information and communication systems and the information contained therein to ensure confidentiality, integrity, and availability. Includes protection and restoration, when needed, of information networks and wire line, wireless, satellite, public safety answering points, and 911 communication systems and control systems.

Defense critical asset—An asset of such extraordinary important to DoD operations in peace, crisis, and war that its incapacitation or destruction would have a very serious, debilitating effect on the ability of DoD to fulfill its mission.

Defense Production Act of 1950 (DPA)—Under Title 1 of the act, as amended (50 U.S.C. App. 2061 et seq.), the president is authorized to require preferential acceptance and performance of contracts or orders supporting certain approved national defense and energy programs, and to allocate materials, services, and facilities in such a manner as to promote these approved programs. Under Title VII, the president is authorized to conduct mandatory surveys and analyses, and prepares reports on specific subsectors of the U.S. DIB. The DPA's current definition of "national defense" includes programs for military and energy production or constructions, military or critical infrastructure assistance to any foreign nation, homeland security, stockpiling, space, and any directly related activity. The DPA authority has also been extended to support emergency preparedness activities under Title VI of the Robert T. Stafford Disaster Relief and Emergency Assistance Act (the Stafford Act), as amended (42 U.S.C 5195 et seq.), and critical infrastructure protection and restoration (Public Law 111-67).

Dependency—The one-directional reliance of an asset, system, network, or collection thereof, within or across sectors, on input, interaction, or other requirement from other sources in order to function properly. A relationship or connection whereby one entity is influenced or controlled by another (Department of Defense Directive [DoDD] 3020.40, 19 Aug 2005).

- **Interdependency**—Relationships or connections between entities of different functions, networks, sectors, or service. (DoDD 30020.40, 19 Aug 2005).
- **Intradependency**—Relationships or connections between entities within a common function, network, sector, or service. (DoDD 3020.40, 19 Aug 2005).

DIB Critical Asset—An asset determined by DoD to be critical to successful execution of its missions.

Government Coordinating Council (GCC)—The government counterpart to the SCC for each sector established to enable interagency coordination. The GCC is comprised of representatives across various levels of government (federal, state, local, tribal, and territorial) as appropriate to the security and operational landscape of each individual sector.

Infrastructure—(1) The framework of networked assets that comprise identifiable industries, institutions, or distribution capabilities that enable continued flow of goods and services (DoDD 3020.40). All building and permanent installations necessary for support, redeployment, and military forces operations (e.g., barracks, headquarters, airfields, communications, facilities, stores, port installations, and maintenance stations) (JP 3-35). (2) The framework of interdependent networks and systems comprising identifiable industries, institutions (include people and procedures), and distribution capabilities that provide a reliable flow of products and services essential to the defense and economic security of the United States, the smooth functioning of government at all levels, and society as a whole. Consistent with the definition in the Homeland Security Act of 2002, infrastructure includes physical, cyber, and/or human elements.

Mitigation—Actions taken in response to a warning or after an incident occurs that are intended to lessen the potentially adverse effects on a given military operation or infrastructure (DoDD 3020.40).

Network—A group or system of interconnected or cooperating entities, normally characteristic as being nodes (assets), and the connections that link them (DoDD 3020.40). The DHS definition of network is a group of components that share information or interact with each other in order to perform a function.

Preparedness—Activities necessary to build, sustain, and improve readiness capabilities to prevent, protect against, respond to, and recover from natural or manmade incidents. Preparedness is a continuous process involving efforts at all levels of government, and between government and the private sector and nongovernmental organizations to identify threats, determine vulnerabilities, and identify required resources to prevent, respond to, and recover from major incidents.

Prioritization—The process of using risk-assessment results to identify where risk-reduction or mitigation efforts are most needed, and subsequently determine which protective actions should be instituted in order to have the greatest effect.

Protection—Preservation of the effectiveness and survivability of mission-related military and nonmilitary personnel, equipment, facilities, information, and infrastructure deployed or located within or outside the boundaries of a given operational area (JP 3-0). DHS states that protection is an action or measure taken to cover or shield from exposure, injury, or destruction. In the context of this book, protection includes actions to deter the threat, mitigate the vulnerabilities, or minimize the consequences associated with a terrorist attack or other incident. Protection can include a wide range of activities, such as hardening facilities, building resilience and redundancy, incorporating hazard resistance into facility design, initiating active or passive countermeasures, installing security systems, promoting workforce surety, training and exercise, and implementing cybersecurity measures, among various others.

Recovery time objective—Target time set for resumption of product, service, or activity delivery after an incident. Note: The recovery time objective must be less than the maximum tolerable period of disruption.

Resilience/resiliency—The adaptive capacity of an organization in a complex and changing environment (ISO 31000).

Note 1: Resilience is the ability of an organization to resist being affected by an event or the ability to return to an acceptable level of performance in an acceptable period after being affected.

Note 2: Resilience is the capability of a system to maintain its functions and structure in the face of internal and external change, and to degrade gracefully when it must.

Note 3: There are three levels of resilience: National, Continuous Integration (CI System), and Individual.

- National—Willpower to bounce back
- CI System—Identify critical nodes for protection and planning prioritization; create adaptive grids to ensure redundancy
- Individual

Response—Activities that address the short-term, direct effects of an incident, including immediate actions to save lives, protect property, and meet basic human needs. Response also includes the execution of emergency operations plans and incident mitigation activities designed to limit the loss of life, personal injury, property damage, and other unfavorable outcomes. As indicated by the situation, response activities include applying intelligence and other information to lessen the effects or consequences of an incident; increasing security operations; continuing investigations into the nature and source of the threat; ongoing surveillance and testing processes; immunizations, isolation, or quarantine; and specific law enforcement operations aimed at preempting, interdicting, or disrupting illegal activity and apprehending actual perpetrators and bringing them to justice.

Risk—Probability and severity of loss linked to threats or hazards (DoDD 3020.40). DHS states risk is the potential for an unwanted outcome resulting from an incident, event, or occurrence, as determined by its likelihood and the associated consequences.

Risk management framework—A planning methodology that outlines the process for setting goals and objectives; identifying assets, systems, and networks; assessing risks; prioritizing and implementing protection programs and resilience strategies; measuring performance; and taking corrective action. Public- and private-sector entities often include risk management frameworks in the business continuity plans.

Sector—A sector is a logical collection of assets, systems, or networks that provide a common function to the economy, government, or society.

System—A functionally, physically, and/or behaviorally related group of regularly interaction or interdependent elements; that group of elements forming a unified whole (JP 3-0) DHS states a system is any combination of facilities, equipment, personnel, procedures, and communications integrated for a specific purpose.

Threat—An adversary having the intent, capability, and opportunity to cause loss or damage (DoDD 3020.40). DHS definition of threat is a natural or man-made occurrence, individual, entity, or action that has or indicates the potential to harm life, information, operations, the environment, and/or property.

Vulnerability—The characteristic of an installation, system, asset, application, or its dependencies that could cause it to suffer a degradation or (incapacity to perform its designated function) as a result of having been subjected to a certain level of threat or hazard (DoDD 3020.40). The DHS definition of vulnerability is a physical feature or operational attribute that renders an entry open to exploitation or susceptible to a given hazard.

REFERENCES AND RECOMMENDED READING

Bergen, P.L. (2002). *Holy War, Inc: Inside the Secret World of Osama bin Ladin*. New York: Touchstone Press, p. 21–22.

CRS (2006). *Chemical Facility Security. CRS Report for Congress*. Washington, DC: Congressional Research Service—The Library of Congress.

DHS (2010). The National Strategy for the Physical Protection of Critical Infrastructures and Key Assets. Accessed at https://www.dhs.gov/xlibrary/assets/physical_Strat.

DHS (2010). Defense Industrial Base Sector-Specific Plan. Washington, DC: U.S. Department of Homeland Security.

FEMA (2015). *Protecting Critical Infrastructure Against Insider Threats.* Accessed 17 April 2015 at http://emilms.fema.gov/IS0915/IABsummary.htm.

Gurr, T.R. (1968). Psychological Factors in Civil Violence. *World Politics*, 20, No. 32, January, 245–78.

Haimes, Y.Y. (2004). *Risk Modeling, Assessment, and Management.* 2nd Edition. New York: John Wiley & Sons, p. 699.

Henry, K. (2002). New Face of Security. *Gov. Security*, Apr., pp. 30–37.

Lindsey, H. (2001). *Vocabulary of Hate.* Accessed 04/18/08 at www.wordnetdaily.com.

NSHS (2006). *National Strategy for Homeland Security.* Accessed 5/13/06 at www .whitehouse/homeland.

New York Times. (1998). Excerpt from court testimony. January 9, 1998, page B4.

Perl, R. (2004). *Terrorism and National Security: Issues and Trends.* CRS Issue Brief IB10119. Washington, DC.

Sauter, M.A., and J.J. Carafano. (2005). *Homeland Security: A Complete Guide to Understanding, Preventing, and Surviving Terrorism.* New York: McGraw-Hill.

Spellman, F.R. (1997). *A Guide to Compliance for Process Safety Management/ Risk Management Planning (PSM/RMP).* Lancaster, PA: Technomic Publishing Company.

The Washington Times (2009). Napolitano Tells It Like It Isn't. Retrieved March 29, 2009 at www.washingtontimes.com/news/2009/mar/29/tell-it-like-it-is-man -casued-disasters-is-mapolit/.

United States Congress (2005). Annual Country Reports on Terrorism. 22 USC, Chapter 38, Section 2656f.

2

Critical Infrastructure

In providing a secure environment, the key function is to connect the dots.

—Frank R. Spellman

So in war, the way is to avoid what is strong and to strike at what is weak.

—Sun Tzu

WHAT IS CRITICAL INFRASTRUCTURE?

For the United States of America, 9/11 was a slap in the face, a punch in the gut (actually, the ultimate sucker punch), and the most serious wake-up call. The 9/11 wake-up call generated several reactions on our part—obviously, protecting ourselves from further attack became (and hopefully still is) priority number one. In light of this important need, the Department of Homeland Security was created. As described by Tom Ridge, the first secretary of the new department, "You may say homeland security is a Y2K problem that doesn't end January 1 of any given year" (Henry 2002). And, according to Barack Obama (2007), the Department of Homeland Security does "the work that ensures no other family members have to lose a loved one to a terrorist who turns a plane into a missile, a terrorist who straps a bomb around her waist and climbs aboard a bus, a terrorist who figures out how to set off a dirty bomb in one of our cities."

Among other things, the new emphasis on homeland security pointed to the need to protect and enhance the security of the nation's critical infrastructure. Critical infrastructure can be defined or listed in many ways. Generally, governments use the term to describe material assets, systems, and services that are essential for the functioning of an economy and society and maintaining public confidence. Destruction or compromise of any of these systems or services would have a debilitating impact on the area either directly, through interdependencies or from cascading effects. For the purpose of this text, critical infrastructure is defined as those assets of key physical resources and computer-/service-based systems that are essential to the minimum operations of economy and government. The 14 critical infrastructures (in author's opinion and making up the volumes in this series) are the following:

- Agriculture
- Financial (banking and finance)
- Chemical and hazardous materials
- Defense industrial base
- Emergency services
- Energy
- Pubic health
- Telecommunications
- Transportation
- Water and wastewater
- Dams
- Information technology
- Commercial and manufacturing assets
- Nuclear power plants

DID YOU KNOW?

Over 85 percent of the critical infrastructure within the United States is owned and operated by the private sector.

Although we did not list digital infrastructure and/or digital operations (e-technology) separately, in this current era we can state without doubt that the digital connection is the glue that holds all critical infrastructure together. This is the case, of course, because all separate infrastructures are interconnected in one way or another. We could go down the list of critical infrastructures and easily point out where e-technology and the specific industry interface. However, we will keep the focus in this text on e-technology as it relates or interfaces with the defense industrial base.

If you're not familiar with the defense industrial base beyond my brief introduction and discussion of it in chapter 1, table 2.1 presents a detailed taxonomy to characterize the DIB sector and help determine critical infrastructure criticality. Note that the characterization provided in table 2.1 primarily applies to impact analysis for DoD mission accomplishment, but also contributes to public health and safety, economic, continuity of government, and public confidence impact analysis. As I pointed out previously, the DIB sector consists of more than 250,000 firms in 215 industries, and it would be impossible to provide a comprehensive accounting of all 250,000 firms and 215 industries. Instead, for illustrative purposes and for a clear and comprehensive discussion, description and explanation of an industry sub-segment, the author has singled out industry segment Mechanical and sub-segment: Bearings. This is why the listing of **BEARINGs** listed in Table 2.1 is highlighted. The Bearing sub-segment will be discussed in chapter 3.

The Defense Industrial Sector provides products and materials that are essential the U.S. defense, offense, security, and economy. The DIB sector is well aware of its vulnerabilities and is leading a significant mandated and voluntary effort to increase its planning and preparedness. Cooperation through industry groups has resulted in substantial information sharing of effective and best practices across the sector. Many sector owners and operators have extensive experience with infrastructure protection and have more recently focused their attention on cybersecurity. In addition to the economic consequences of a successful home-grown or foreign terrorist attack against DIB sector facilities, there is also the potential of a threat to public health and safety and the environment.

Table 2.1. DIB Segments, Sub-Segments, and Communities

Industry Segments	
Industry Segments	*Industry Sub-Segment*
Aircraft	Fixed Wing
	Rotary Wing
	Unmanned Aerial Systems
Ships	Surface
	Sub-Surface
	Unmanned Underwater Vehicles
Tracked and Wheeled Land Vehicles	Combat Vehicles
	Tactical Vehicles
	Unmanned Ground Vehicles
Electronics	Electronic Warfare
	Command, Control, Communications, Computer and Intelligence
	Avionics
Soldier Systems	Chemical Biological Defense Systems
	Clothing and Textiles
	Subsistence/Medical
Structural	Castings/Forgings
	Composites
	Armor (Ceramic/Plating)
	Precious Metals
Munitions	Missile Tactical
	Missile Strategic
	Missile Air/Air
	Missile Air/Surface
	Missile Defense
	Missile Surface/Air
	Missile Surface/Surface
	Precision Guided Munitions
	Ammunition
	Missile Defense Agency
Space	Launch Vehicles
	Satellite
	Missile Defense Agency
Mechanical	Transmissions (Air/Auto)
	Propulsion (Diesel/Rocket/Turbine)
	Hydraulics
	BEARINGS
	Nuclear Components (including depleted Uranium)

REFERENCES AND RECOMMENDED READING

CBO (2004). *Homeland Security and the Private Sector*. Washington, DC: Congressional Budget Office.

DHS (2003). The National Strategy for the Physical Protection of Critical Infrastructures and Key Assets. Accessed at https://www.dhs.gov/xlibrary/assets/physical_Strat.

DHS (2010). Defense Industrial Base Sector-Specific Plan. Washington, DC: U.S. Department of Homeland Security.

FAO (2005). *Bridging the Rural Digital Divide*. New York: United Nations. Accessed 04/19/08 at www.fao.org/rdd.

FEMA (2015). *Protecting Critical Infrastructure Against Insider Threats.* Accessed 17 April 2015 at http://emilms.fema.gov/IS0915/IABsummary.htm.

FR (2003). Notice of Proposed Rulemaking. *Federal Register*, vol. 68, no. 90.

Henry, K. (2002). New Face of Security. *Gov. Security*, Apr., pp. 30–37.

Horn, F.P. (1999). Statement made before the United States Senate Emerging Threats and Capabilities Subcommittee of the Armed Services Committee. Accessed 06/27/07 at www.Senate.gov/~armed_servies/statement/1999/991027fh.pdf.

Lane, J. (2002). Sworn Testimony, Congressional Field Hearing, House Committee on Government Reform, Abilene, KS.

Obama, B. (2007). *Homeland Security*. http://www.whitehouse.gov/agenda/home land_security/.

3

Have a Bearing in Mind?

In the prevention of and preparation for terrorism, ending well is the best revenge.

—Frank R. Spellman

The flow of providing security protection, from data to understanding:

DATA → INFORMATION KNOWLEDGE → UNDERSTANDING

GETTING YOUR BEARINGS[1]

The sliding of a heavy load or object using sledges came before the rolling of an object using some type of rollers or wheels, but the advantages of rolling eventually became apparent. And the development of roller bearings evolved from experiments involving intuitive common sense and practical applica-

1. The material presented in this chapter is based on F.R. Spellman (2016) *Conveyance of Water/Wastewater*. Boca Raton, FL: CRC Press; USDOC (1993) National Security Assessment of the Antifriction Bearings Industry. Washington, DC: U.S. Department of Commerce.

tions. From the development of the wheel sprang the need to reduce friction caused by the wheel turning on some type of axle. Thus, because of the need to reduce friction, emerged the anti-friction roller bearing.

The bearing is important, but if you ever think of the word *bearing*—or hear it in conversation—you're apt to define it as it relates to manners—behavior, attitude, air, comportment, and so on—or in terms of a compass reading—direction, course, orientation, a point of reference—or maybe you associate the word with influence—impact, connection, relevance, or sway. Well, be advised that none of these definitions of *bearing* is what this chapter is about! Instead, we should think about bearings in regard to the movement of heavy stone building blocks and carvings, road vehicles, precision instruments, water-raising equipment, and windmills. Accordingly, the bearing we focus on here is usually qualified by a preceding descriptor such as roller, rolling, sleeve (figure 3.1), ball, needle, tapered and thrust, and others. Because this is a book about homeland security and the protection of life, limb, and property from terrorist acts of

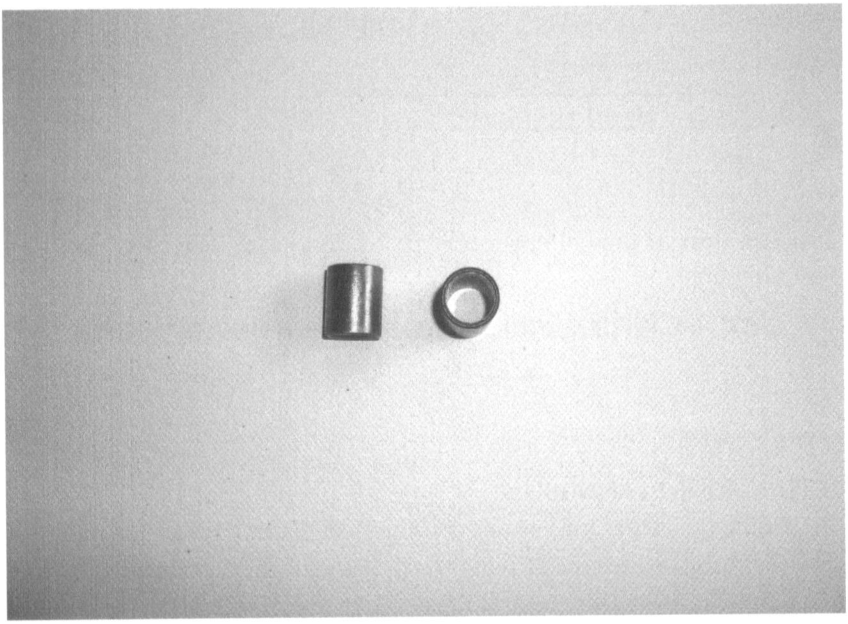

FIGURE 3.1
Sleeve Bearings

destruction you might think that this chapter is about those bearings (among other objects) that terrorists place in improvised explosive devices (IEDs). Those assorted bearings are designed for one purpose: death, fear, and destruction. This chapter is not about using bearings in weapons, but instead is about terrorist activities designed to destroy manufacturing activities that produce bearings that are used in DIB machinery, equipment, and weapons.

NOT THOUGHT OF UNTIL IT FAILS OR IS NEEDED

Now that we have qualified the type of bearings we are concerned with here in this discussion, it is important to point out that even though they play an important role not only in our national defense but also our daily lives, roller, rolling, sleeve, ball, needle, tapered, and thrust bearings are not commonly given a first or second thought by most of us. However, when you hear your home washing machine run sounding like it is full of marbles or when the axle of your automobile smokes, burns, fuses, and stops allowing a tire to rotate, you quickly become acquainted with roller bearings of one type or another.

And then there are the extreme failures that put roller bearings right square in the mouth, so to speak. Imagine your skilled dentist drilling away on one of your teeth when the bearings in the drill fail. Ouch! Or, think about this—you are driving down the road and the car's water pump freezes up and fails because of bearing failure. Keep in mind that there is almost no critical operation where rolling bearings are not used, especially throughout the durables goods sector to reduce friction. Along with being used in washers and dryers, markets are as varied as automobiles, airplanes machine tools, computers, conveyors, construction equipment, elevators, or virtually any item have rotating parts. And their importance in daily activities becomes clear—even for a short moment of time (until the washer is replaced or car repaired).

Rolling bearings are of critical important in defense applications. Put yourself in the place of a fighter pilot, military tank turret operator, naval ship commander, or any warrior using machines to combat an enemy in the field, on the seas, in the air, in space, or wherever and the importance of roller bearings is not only apparent, but also could be life-saving—for the operator. How successful would combatants be if their war-fighting apparatus, such as missile guidance systems, fails due to bearing failure?

DID YOU KNOW?

Roller bearings provide maintenance of a very low friction coefficient over a wide range of specified rotational speeds, significantly reducing energy requirements and greatly increasing machine efficiency and precision.

And, again, unless you are a maintenance person, one who deals with maintenance activities like checking on (performing preventive maintenance) or replacing defective bearings with new bearings, bearings are probably one of the furthest items from anyone's mind. When your lawnmower, weed-wacker, handheld drill motor, or any rotating or reciprocating hand tool freezes up, you will know what bearing failure is.

WHAT ARE ROLLER BEARINGS?

Rolling (or antifriction) bearings are mass-produced, high-precision elements. They consist of rolling components sandwiched between inner and outer rings (see figure 3.2). The roller elements include balls (see figure 3.3), cylindrical rollers, barrel or spherical shapes, tapers, and elongated needle-like rollers. It is the type of rolling element employed that gives the rolling bearing its name. The outer ring is usually attached to a machine's structure, like in a pump, for example. On the other hand, the inner ring (the bore) is attached to a rotating shaft so that the rolling elements roll in the opposite direction of the shaft. The load is carried or transmitted across the rolling elements. Figure 3.4 not only shows how bearings are used in a centrifugal pump but also points to their importance in the operation of the pump. More importantly, from figure 3.4 it can be seen that a failure to any of the centrifugal pumps bearings would be detrimental to its operation. Figure 3.4 also shows that a centrifugal pump is a precision machine. Thus, to operate at design levels its components must be at precision levels. In many types of machines using bearings, precision levels to the millionths of an inch are not uncommon and are critical to reduce noise, vibration, heat, and bearing wear.

FIGURE 3.2
Inner and Outer Ring of Bearing

FIGURE 3.3
Roller Bearings and Ball Bearings

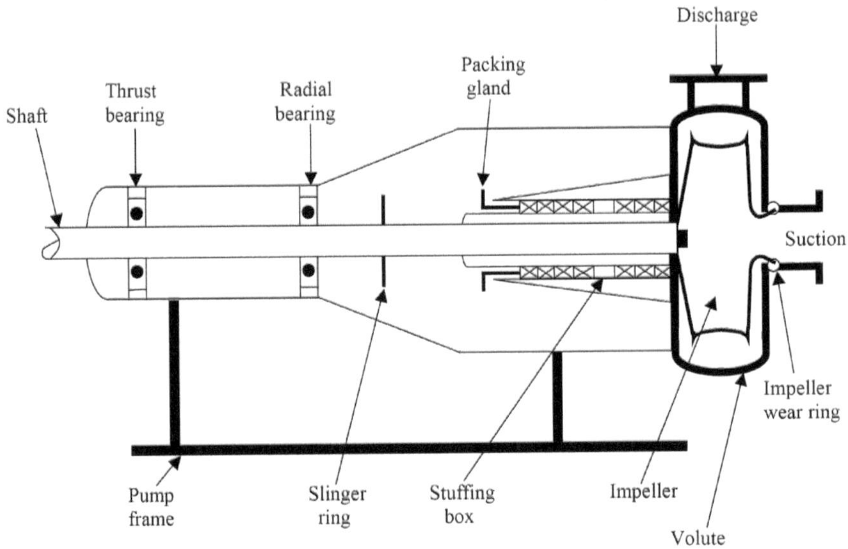

FIGURE 3.4
Major Parts of a Centrifugal Pump

If you were to look at a catalog listing of bearing part numbers used worldwide, you would probably be surprised to find a telephone book-sized collection of part numbers—more than 150,000 distinct bearings would be listed—although many of the differences are rather superficial. Actually, close scrutiny of bearing catalogs shows that only about two thousand bearing part numbers are manufactured in high volume, and these are the most actively traded across international borders. Bearing sizes range from as small as only 1 millimeter in diameter to nearly 40 feet across; however, the great majority of them lie between about 0.5 and 8 inches.

Rolling bearings are identified in relation to the type of rolling element they contain as well as by their size and precision level. High-quality, super-precision bearings tend to be the most problematic and generally have the longest lead times, although larger-sized regular-precision ones can cause problems as well. Ball bearings have the least friction of the bearing types because of the "point" contact between the rolling element and raceway surfaces. They also have the lowest per unit price because they are the easiest to make. Roller bearings have "linear" contact at the rolling surfaces, which sacrifices speed for great load carrying capability. Roller bearings, with the exception of needle bearings, tend to be larger and much more expensive than ball bearings. Roller bearings are normally the choice for larger equipment—trucks, mining equipment, forging presses, and so on.

Types and Applications of Ball and Roller Bearings

Some typical types and applications of ball and roller bearings include the following:

- *Miniature and instrument bearings* (i.e., radial ball bearings under 30 mm in outside diameter)—Small electric motors, computers, photocopy equipment, communication equipment, power hand tools, gyroscopes, dental drills, instrumentation.
- *Single-row deep-groove ball bearings* (i.e., radial ball bearings generally over 30 mm and under 200 mm in outside diameter)—Centrifugal pumps, conveyors, electric motors, generators, machine tools, automobile and truck components.
- *Maximum capacity ball bearings* (i.e., ball bearings with the maximum compliment of rolling elements, can carry more load)—Industrial lift trucks, construction machinery, truck and farm equipment components, heavy duty pumps, gear reducers.
- *Angular contact ball bearings* (i.e., non-radial ball bearings that counter axial thrust in one direction; often mounted in opposing pairs)—Gear drive clutch release, compressors, wheel hubs, printing presses, pumps, electric motors, machine tools.
- *Double-row ball bearings* (i.e., radial ball bearings with two rows that distribute load over a wider area; able to carry more load)—Automotive water pumps, compressors, automotive pinions, power transmissions.
- *Integral shaft ball bearing* (i.e., typically a double-row radial ball bearing with the inner raceways built into the rotating shaft; saves pace)—Automotive water pumps, compressors, automatic pinions, power transmissions.
- *Self-aligning ball bearings* (i.e., ball bearings whose outer ring raceway structure allows lateral displacement; needed where rotating shafts can bend under stress)—Farm equipment, power transmissions, heavier equipment.
- *Thrust ball bearings* (i.e., non-radial ball bearings with raceways perpendicular to axis of rotation, oppose axial thrust in both directions; prevent axial movement)—Machine tools, three-wheeled vehicles, valves, cranes tapping machines, mining machines.

- *Cylindrical roller bearings* (i.e., rollers width and length about equal; carries radial load)—Rolling mills, machine tools, electric motors, printing machinery, pumps, oil field machinery, railway equipment.
- *Needle bearings* (i.e., rollers length four times or more the width; can carry very high radial load in small diameters)—Automotive steering columns, planetary gears, machine tools, printing machinery, air compressors, textile machinery, woodworking machinery, gear drives, universal joints.
- *Tapered roller bearings* (i.e., rollers tapered or conical, designed to carry radial and thrust loads, often installed in opposing pairs)—Wheel units, automobiles, differentials, tracks, mail equipment, machine tools, railroad wheel boxes, farm machinery, milling machines, printing machinery, textile machinery, turret lathers, valves, trailers, foundry equipment.
- *Spherical roller bearings* (i.e., rollers arced along width that self-align with rotating, bending shaft under extreme stress; can carry very large loads)—Construction equipment, crushers, power shovels, steel rolling mills, pulverizers, textile machinery, mining equipment

THE LAST WORD ON THE IMPORTANCE OF ROLLER BEARINGS

How important is the roller bearing industry to our national defense? Well, consider that in August of 1988, the Office of the Secretary of Defense issued a Defense Federal Acquisition Regulation (DFAR) that required that all Department of Defense (DoD) purchases of antifriction bearings (except ball bearings under 30 mm or 1.181 inches outside diameter) be limited to U.S. or Canadian sources. The DFAR was established to address DOD concerns regarding the erosion of this defense-critical industry (USDOC 1993).

Many bearing manufacturers reported that the DFAR had a positive impact on their production capacity, employment, investment, R & D, and profitability. Moreover, the manufacturers commented that the DFAAR improved entrée to defense prime contractors, increased awareness of U.S. bearing producers' capabilities, and supported U.S. maintenance of technological proficiency in super-precision bearing production.

The bottom line: any of the DIB critical infrastructure segments or subsegments shown in table 2.1 could have been selected as the focus of this chapter. However, because ball bearings are also one of the favorite ingredients in terrorist bombs to maim and kill innocents, it seemed appropriate to point out that the more important intended purpose of roller bearings are much more far-reaching, important, and human-friendly.

REFERENCES AND RECOMMENDED READING

Bond, C. (2002). Statement on S. 2579. *Congressional Record,* Daily Edition.

DHS (2003). *The National Strategy for the Physical Protection of Critical Infrastructure and Key Assets.* Washington, DC.

DHS (2008). Homeland Security Presidential Directives. Accessed at www.dhs.gov/xabout/laws/editorial_0607.shtm.

FEMA (2015). *Protecting Critical Infrastructure Against Insider Threats.* Accessed 17 April 2015 at http://emilms.fema.gov/IS0915/IABsummary.htm.

NASA (1981). *History of Ball Bearings.* Washington, DC: NASATM-81689 198110009866.

USDOC (1993). *National Security Assessment of the Antifriction Bearings Industry.* Washington, DC: United States Department of commerce.

USDOE (2009). *Science and Technology.* Retrieved 04/25/09 athttp://www.energy.gov/ sciencetech/index.htm.

4

Defense Industrial Base Security

The emergence of amorphous and largely unknown terrorist individuals and groups operating independently (freelancers) and the new recruitment patterns of some groups, such as recruiting suicide commandos, female and child terrorists, and scientists capable of developing weapons of mass destruction, provide a measure of urgency to increasing our understanding of the psychological and sociological dynamics of terrorist groups and individuals.

—Hudson

The only truly secure system is one that is powered off, cast in a block of concrete, and sealed in a lead-lined room with armed guards.

—Gene Spafford

WHY THE DEFENSE INDUSTRIAL BASE SECTOR?[1]

While it is true that the U.S. Defense Industrial Base (DIB) sector has not been attacked on a large scale by terrorists (as far as we know and have been told), it is also true that the potential risk is more than conjecture. From several terrorism incidents on critical infrastructure in South America, the Middle East, and Africa, it is apparent that DIB infrastructure is a preferred target of

1. The information contained in this chapter is based on DHS (2007) *Defense Industrial Base Sector—Sector Plan.* Accessed at http://www.dhs.gov/xlibrary/assets/nipp-spp-defense.

the terrorists. Moreover, Al Qaeda has made it clear that it has an interest in attacking the American DIB system.

The obvious question might be: What makes the DIB sector such an attractive terrorism target? The DIB sector is an attractive target because of the following:

- DIB sector components or assets are spread throughout the nation with little definition of boundaries.
- Much of the DIB sector equipment is one-of-a-kind large machinery which is difficult to replace in the short term.
- Several DIB sector components were designed and constructed without concern for terrorist intrusion or destructive activities.
- Many DIB systems are monitored and operated using underprotected SCADA systems.
- As with many of industries attempting to economize, many DIB segment and subsegments assign responsibility for safety and security as a collateral duty to a line employee instead of employing a fulltime certified safety and security professional.

VISION STATEMENT FOR THE DEFENSE INDUSTRIAL BASE SECTOR

Ensure the ability of the DIB to support DoD missions and eliminate unacceptable risk to national security through informed infrastructure risk-management decisions.

DIB SECTOR SECURITY GOALS AND ATTRIBUTES

The Department of Defense has identified eight general elements and characteristics of sector security goals and attributes.

- **Critical Asset Reduction Goal.** Sector resiliency will be most assured if no particular asset can be assessed as more critical than any other. While the ultimate ideal goal would be zero-critical DIB assets, the sector will strive to reduce the number of critical assets whenever and wherever possible within fiscal and legal constraints. Sound risk management practices, including asset resiliency, mitigation of risks, and redundancy will be shared and advanced throughout the sector.

- **Cyber Goals and Attributes**. Like physical attributes, these assist the DIB sector to evaluate consequences and vulnerabilities, and develop protective strategies. Cyber systems that link and help monitor and control the DIB systems are increasingly recognized as a potential vulnerability. All information that identifies or otherwise describes characteristics of a critical DIB asset that is created, held, and maintained by the government or the private sector will be protected from unauthorized disclosure according to established procedures, appropriate to the particular level of information.
- **Volumetric or Throughput Attributes**. These define the extent of any damage, depending on the utilized capacity of the systems, or points where the system may be capacity constrained.
- **Personnel Security Goals and Human Attributes**. Ensure all personnel directly associated with a critical DIB asset are vetted for employment suitability, reliability, and trustworthiness using established processes commensurate with requirements of the respective positions held, in conformance with pertinent security policy. Highly trained and skilled personnel are key factors in a comprehensive DIB sector security plan. The availability of skilled and experienced technical talent is a concern in the DIB sector. Sustaining essential technical knowledge a critical to maintaining the sector's safety, reliability, and security.
- **Physical Security Goal.** Determine the impact or consequence of critical DIB asset loss to the DoD mission(s) supported, the known or perceived threat, and the susceptibility to exploitation of vulnerabilities the threat is capable of perpetuating; identify specific DIB assets the destruction or disruption of which could result in human casualties or economic disruption similar to the effects of weapons of mass destruction; compile a composite of facility physical security risk assessments.
- **Insider Threat Goal and Attributes.** DoD provides security education and training aids to DIB asset owners/operators not having security program so that they may implement provisions for the vetting of system and network administrators commensurate with the consequences of the loss of sensitive

or classified information, production or provisioning capability, and supply chain integrity.

- **Monitoring and Reporting Goals and Attributes.** On-going determination of the effectiveness of government threat reporting to officials, owners, and operators responsible for critical DIB assets, and to local law enforcement officials and other first-responders including, as appropriate, the medical and mass transportation communities.
- **Training and Education Goal and Attributes**. Develop and provide continuous specific security education and training materials for critical DIB asset owner/operators.

HOMELAND SECURITY DIRECTIVES

As may be obvious, Homeland Security Presidential Directives are issued by the president. Each directive has specific meaning and purpose and is carried out by the U.S. Department of Homeland Security. Table 4.1 lists Homeland Security Presidential Directives (HSPDs).

ASSESSING CONSEQUENCES

The potential physical and cyber consequences of any incident, including terror attacks and natural or human-made disasters, are the primary consideration in risk assessment. In the context of this text, consequence is measured as the range of loss or damage that can be expected. The consequences that are considered for the national-level comparative risk assessment are based on the criteria as set forth in HSPD-7. These criteria can be divided into four main categories:

- **Human Impact:** Effect on human life and physical well-being (e.g., fatalities, injures).
- **Economic Impact**: Direct and indirect effects on the economy (e.g., costs resulting from disruption of products or services, costs to respond to and recover from the disruption, costs to rebuild the asset, and long-term costs due to environmental damage).
- **Impact on Public Confidence**: Effect on public morale and confidence in national economic and political institutions.
- **Impact on Government Capability**: Effect on the government's ability to maintain order, deliver minimum essential public services, ensure public health and safety, and carry out national security-related missions.

Table 4.1. Homeland Security Presidential Directives

HSPD – 1: Organization and Operation of the Homeland Security Council. (White House) Ensures coordination of all homeland security-related activities among executive departments and agencies and promotes the effective development and implementation of all homeland security policies.
HSPD – 2: Combating Terrorism through Immigration Policies. (White House) Provides for the creation of task force which will work aggressively to prevent aliens who engage in or support terrorist activity from entering the United States and to detain, prosecute, or deport any such aliens who are within the United States.
HSPD – 3: Homeland Security Advisory System. (White House) Establishes a comprehensive and effective means to disseminate information regarding the risk of terrorist acts to federal, state, and local authorities and to the American people.
HSPD – 4: National Strategy to Combat Weapons of Mass Destruction. Applies new technologies, increases emphasis on intelligence collection and analysis, strengthens alliance relationships, and establishes new partnerships with former adversaries to counter this threat in all of its dimensions.
HSPD – 5: Management of Domestic Incidents. (White House) Enhances the ability of the United States to manage domestic incidents by establishing a single, comprehensive national incident management system.
HSPD – 6: Integration and Use of Screening Information. (White House) Provides for the establishment of the Terrorist Threat Integration Center.
HSPD – 7: Critical Infrastructure Identification, Prioritization, and Protection. (White House) Establishes a national policy for federal departments and agencies to identify and prioritize United States critical infrastructure and key resources and to protect them from terrorist attacks.
HSPD – 8: National Preparedness. (White House) Identifies steps for improved coordination in response to incidents. This directive describes the way federal departments and agencies will prepare for such a response, including prevention activities during the early stages of a terrorism incident. This directive is a companion to HSPD-5.
HSPD – 8 Annex 1: National Planning. Further enhances the preparedness of the United States by formally establishing a standard and comprehensive approach to national planning.
HSPD – 9: Defense of United States Agriculture and Food. (White House) Establishes a national policy to defend the agriculture and food system against terrorist attacks, major disasters, and other emergencies.
HSPD – 10: Biodefense for the 21st Century. (White House) Provides a comprehensive framework for our nation's biodefense.
HSPD – 11: Comprehensive Terrorist-Related Screening Procedures. (White House) Implements a coordinated and comprehensive approach to terrorist-related screening that supports homeland security, at home and abroad. This directive builds upon HSPD – 6.
HSPD – 12: Policy for a Common Identification Standard for Federal Employees and Contractors. (White House) Establishes a mandatory, government-wide standard for secure and reliable forms of identification issued by the federal government to its employees and contractors (including contractor employees).
HSPD – 13: Maritime Security Policy. Establishes policy guidelines to enhance national and homeland security by protecting U.S. maritime interests.
HSPD—14: Domestic Nuclear Detection establishes a Domestic Nuclear Detection Office (DNSO) to coordinate efforts to protect the domestic United States against dangers from nuclear or radiological materials. EPA supports the detection, response, law enforcement, and information sharing aspects of the DNDO's mission.
HSPD – 16: Aviation Strategy. Details a strategic vision or aviation security while recognizing ongoing efforts, and directs the production of a National Strategy for Aviation Security and supporting plans.
HSPD – 18: Medical Countermeasures against Weapons of Mass Destruction. (White House) Establishes policy guidelines to draw upon the considerable potential of the scientific community in the public and private sectors to address medical countermeasure requirements relating to CBRN threats.
HSPD – 19: Combating Terrorist Use of Explosives in the United States. (White House) Establishes a national policy, and calls for the development of a national strategy and implementation plan, on the prevention and detection of, protection against, and response to terrorist use of explosives in the United States.
HSPD – 20: National Continuity Policy. (White House) Establishes a comprehensive national policy on the continuity of federal government structures and operations and a single National Continuity Coordinator responsible for coordinating the development and implementation of federal continuity policies.
HSPD – 21: Public Health and Medical Preparedness. (White House) Establishes a national strategy that will enable a level of public health and medical preparedness sufficient to address a range of possible disasters.
HSPD—23: Cybersecurity requires federal agencies to monitor cyber activity toward federal agencies' computer systems and where necessary, provide action to eliminate sources of hostile action. EPA has a robust security program for both personnel and cybersecurity as mandate by the directive.

Source: USEPA (2016).

An assessment of all categories of consequence may be beyond the capabilities available for a given risk analysis. Most DIB sector assets are not associated with the possibility of mass causalities, but may have economic and long-term health and safety consequences.

REFERENCES AND RECOMMENDED READING

Crayton, J.W. (1983). Terrorism and the Psychology of the Self, pages 33–41 in Lawrence Zelic Freedman and Yonah Alexander, eds., *Perspectives on Terrorism*. Wilmington, Delaware: Scholarly Resources.

DHS (2007) *Defense Industrial Base Sector—Sector Plan*. Accessed at http://www.dhs.gov/xlibrary/assets/nipp-spp-defense.

FEMA (2015). *Protecting Critical Infrastructure Against Insider Threats*. Accessed 17 April 2015 at http://emilms.fema.gov/IS0915/IABsummary.htm.

Ferracuti, F. (1982). A Sociopsychiatric Interpretation of Terrorism, *The Annals of the American Academy of Political and Social Science*, 463, September, 129–41.

FR (2007). *Federal Register* 17688–17745.

Hudson, R.A. (1999). *The Sociology and Psychology of Terrorism: Who Becomes a Terrorist and Why*? Washington, DC: Library of Congress.

Lees, Frank (1996). *Loss Prevention in the Process Industries* 3:A5.1–A5.11. New York: Butterworth-Heinemann.

Long, D.E. (1990). *The Anatomy of Terrorism*. New York: Free Press.

Margolin, J. (1977). Psychological Perspectives on Terrorism. In Y. Alexander and S.M. Finger (eds.). *Terrorism: Interdisciplinary Perspectives*. NY: John Jay Press.

OMB (1998). *Federal Conformity Assessment Activities, Circular A-119*. Washington, DC: White House.

Pearlstein, R. (1991). *The Mind of the Political Terrorist*. Wilmington, Del: Scholarly Resources, Inc.

Spafford, G. (1989). In Computer Relations: Of Works, Viruses, and Core War. A.K. Dewdney. *Scientific American*. March 1989: p 110.

Sullivant, J. (2007). *Strategies for Protecting National Critical Infrastructure Assets: A Focus on Problem-Solving*. New York: Wiley & Sons.

USEPA (2016). *Homeland Security Presidential Directives*. Accessed at https://www.epa.gov/emeergency-resposne/horelamnd-security-presidential-directives.

5

DIB Sector Vulnerability Assessment

But there's one thing we must all be clear about: terrorism is not the pursuit of legitimate goals by some sort of legitimate means. Whatever the murderers may be trying to achieve, creating a better world certainly isn't one of their goals. Instead they are out to murder innocent people.

—Salman Rushdie

The DIB sector is classified as "vulnerable" in the sense that inherent weaknesses in its operating environment could be exploited to cause harm to the system. There is also the possibility of a cascading effect—a chain of events—due to a terrorist act affecting DIB sector providers, which could cause corresponding damage (collateral damage) to other nearby users. In addition to significant damage to the nation's DIB sector, entities using and needing DIB-produced materials to function result in loss of life due to a lack of proper supplies; shutdown of other industries; catastrophic environmental damage to rivers, lakes and wetlands; and long-term public health impacts.

ASSESSING VULNERABILITIES

For the purpose of this text and according to FEMA 2008, vulnerability is defined as any weakness that can be exploited by an aggressor to make an asset susceptible to hazard damage.

In addition, according to the Department of Homeland Security (DHS 2009), vulnerabilities are physical features or operational attributes that render an entity open to exploitation or susceptible to a given hazard. Vulnerabilities may be associated with physical (e.g., a broken fence), cyber (e.g., lack of a firewall), or a human (e.g., untrained guards) factors.

Conducting vulnerability assessments is conducted on many levels. DoD has instituted a process to provide awareness training to DIB asset owner/operators. The purpose of the awareness training is to provide DIB company personnel with information about the place of their asset within the overall DoD mission requirements and acquisition process so they will understand their rules and importance to the entities at the corporate and site levels (DoD 2010). The training focuses on:

- Protection of DoD interests
- Protection of federal interests
- Mission assurance to the war fighters
- Importance of facilities fostering relationships with local responders and federal, state, and local law enforcement/civil authorities for business recovery planning.

The awareness training also informs the asset owner/operators of the protection measures applied to their proprietary and business-sensitive information provided to DoD, as well as an overview of the DoD, as well as an overview of the DoD information-sharing responsibilities and procedures regarding this data.

Once critical DIB assets are identified and prioritized, the next step is to conduct standardized assessments. DoD, working though Defense Contract Management Agency (DOMA) and the National Guard Bureau, has established a standardized mission assurance assessment for application to critical DIB assets. These assessments consider impact, vulnerability, and threat/hazard (whether from natural disaster, technological failure, human error, criminal activity, or terrorist attack). This approach to risk assessment ensures consideration of relevant factors for each DIB asset and a relative prioritization of risks to support military operations. National Guard Mission Assurance Assessment teams have been designated to conduct assessments of DIB assets in the United States.

INSIDER THREAT VULNERABILITY

An important part of any successful vulnerability assessment process is awareness. Security and risk managers (and all employees in general) working in or with the critical infrastructure sectors must be aware of the potential for insider threat vulnerability. Again, the key word is AWARENESS. Awareness means that personnel within the critical infrastructure sectors must know how to identify and take action against insider threats. To achieve this critical goal, safety and security personnel must be provided with an overview of and be cognizant of common characteristics and indicators associated with malicious insiders and effective measures to counter insider threats.

Protecting Against Insider Threats (FEMA 2015)

As mentioned earlier, when analyzing threats to our nation's critical infrastructure, we tend to focus on malicious actions from boat- or planeloads of outside actors. Of equal concern (and even more so in the author's view) are threats from an insider—someone to whom we have given legitimate access to information, systems, and resources. The measures we take to detect and protect against external threats may not be sufficient to address threats from insiders.

A malicious insider has access and inside knowledge of the organization and uses that knowledge with the intent to cause harm. The insider may be a current employee, a former employee, a service provider, or, especially in the current era, a planted person inside who has been brainwashed and turned into a terrorist ("radicalized") waiting for the moment of maximum impact to people and property.

Given the importance of our nation's critical infrastructure, the actions taken by a malicious employee or service proceed could have devastating consequences. Let's look at some actual examples.

- In June 2015, at a U.S.-owned gas factory in Lyons, France, an employee citing allegiance to ISIS beheaded his supervisor and rammed a truck into the facility to cause an explosion.
- A service provider employee at a nuclear facility stole two five-gallon containers of low-enriched uranium dioxide and then attempted to extort $100,000 by threatening to disperse the material in an unnamed U.S. city.

- A power company field engineer, angry with his supervisor, disabled protection systems at a substation and forced the shutdown of the entire network. More than 800,000 customers lost power as a result.
- Two municipal employees used their access credentials to sabotage the system controlling the traffic lights of a major city, causing widespread traffic delays. The damage took four days to repair.

Insider threats endanger the integrity and security of our workplaces and our communities. This section helps you become aware of threat indicators and actions you can take.

The Insider Threat Defined

The President's National Infrastructure Advisory Council (2008) defines the insider threat as follows:

> The insider threat to critical infrastructure is one or more individuals with the access or inside knowledge of a company, organization, or enterprise that would allow them to exploit the vulnerabilities of the entity's security, systems, services, products, or facilities with the intent to cause harm.

A person who takes advantage of access or inside knowledge in such a manner commonly is referred to as a "malicious insider."

The Scope of Insider Threats

Insider threats can be accomplished through either physical or cyber means and may involve any of the following:

- **Physical or information-technology sabotage**—Involves modification or damage to an organization's facilities, property, assets, inventory, or systems with the purpose of harming or threatening harm to an individual, the organization, or the organization's operations.
- **Theft of intellectual property**—Involves removal or transfer of an organization's intellectual property outside the organization through physical or electronic means (also known as economic espionage).
- **Theft or economic fraud**—Involves acquisition of an organization's financial or other assets through theft or fraud.

DID YOU KNOW?

The FBI testified in June 2012 that in the preceding year, economic espionage losses to the American economy totaled more than $13 billion. In the previous four years, the number of arrests the FBI made had doubled; indictments increased by a factor of five; and convictions increased by a factor of eight (FBI 2012). In another survey, security professionals found that 43.2 percent of respondents attributed some loss at their organization to insiders. Forty-six percent of respondents said the damage caused by insider attacks was more damaging than outside attacks (CSI 2011).

- **National security espionage**—Involves obtaining information or assets with a potential impact on national security through clandestine activities.

Insider Threats to DIB Sector

Insider threats to a critical infrastructure can occur over many years and have devastating consequences. Malicious incidents have occurred throughout the DIB infrastructure sector even though the penalties are severe: under the provisions of the Economic Espionage Act of 1996, penalties can include fines of up to $500,000 per offense and imprisonment of up to fifteen years for individuals, as well as criminal forfeiture of any proceeds of the crime. The fact is no fine or other penalty is going to deter a dedicated insider terrorist from performing his or her evil act(s).

One of the most insidious, ongoing, and unnoticed examples, for months or even years, of an increasing insider threat is intellectual property theft. Intellectual property theft can seriously compromise our ability to protect ourselves from outside attack, as shown in the examples below:

- Greg Chung was a Boeing engineer who, from 1979 to 2006, stole trade secrets about the space shuttle, the Delta IV rocket, and the C-17 military cargo jet. He sent that sensitive information to China, his native country. Chung was sentenced in February 2010 to more than fifteen years in prison.

- The supervisor of a maintenance crew with access to every office at a commercial research and development facility stole confidential research information and tried to sell it to a rival company. He pled guilty in 1997 to theft of trade secrets and was sentenced to fifteen months in jail.

Common Characteristics and Traits of Malicious Insiders

Based on research conducted by the Software Engineering Institute at Carnegie Mellon University and the U.S. Secret Service National Threat Assessment Center, malicious insiders often are perceived or known to be difficult or high-maintenance employees who are:

- Obviously unhappy or extremely resentful
- Having financial, performance, or behavioral problems
- At risk (or perceived to be) for layoff or termination

Keep in mind that not all malicious insiders fit this characterization. Insiders involved in national security espionage, for example, may exhibit few outward signs. In the majority of cases, however, management and/or human resources personnel were well aware of the employees and their issues prior to an incident.

The following motives and personal situations frequently are linked with malicious insiders:

- **Personal or Behavioral Problems**
  - Vulnerable to blackmail
  - Experiencing family or financial problems
  - Prone to compulsive or destructive behavior
  - Subject to ego or self-image issues
- **Personal Desires**
  - Seeking adventure or thrill
  - Seeking approval and returned favors
  - Professing allegiance

- **Workplace Issues**
    - Experiencing problems at work
    - Feeling anger or need for revenge

Organizational Factors That Embolden Malicious Insiders

The following organizational factors have been known to encourage or present opportunities to potential malicious insiders.

- **Access and Availability**
    - Ease of access to materials and information
    - Ability to exit the facility or network with materials or information
- **Policies and Procedures**
    - Undefined or inadequate policies and procedures
    - Inadequate training
    - Lack of training
- **Time Pressure and Consequences**
    - Rushed employees
    - Perception of lack of consequences

Insider Activities and Behavior You May See

Insider threats may be detected through particular activities and behavior on the part of the insider. These activities and behaviors often will appear unusual or suspicious. Keep in mind there may be several explanations for a particular activity or behavior identified here, but when combined with other factors, certain activity or behaviors point toward a possible insider threat. A combination or confluence of indicators should not be ignored.

Unusual or suspicious insider activities and behavior can be described using the following categories:

- Inappropriate interest or acquisition
- Unauthorized or unusual computer use
- Unusual hours, contacts, or travel

- Secretive or peculiar behavior
- Personal or financial issues

So what can managers, supervisors, or employers do? There are a range of preventive actions that can be taken:

- Using appropriate screening processes to select new employees
- Becoming familiar with behavior and activities associated with malicious insiders
- Clearly communicating and consistently enforcing security policies and controls
- Encouraging and providing non-threatening, convenient ways for employees to report suspicious in a confidential manner
- Ensuring that proprietary information and materials are adequately, if not robustly, protected
- Routinely monitoring computer networks for suspicious activity
- Ensuring security (to include computer network security) personnel have the tools they need
- Consulting with legal and law enforcement experts as needed to ensure compliance with the law
- Documenting and evaluating incidents of suspicious or disruptive behavior

VULNERABILITY ASSESSMENT[1]

A **vulnerability assessment (VA)** involves an in-depth analysis of the facility's functions, systems, and site characteristics to identify weaknesses and lack of redundancy and to determine mitigations or corrective actions that can be designed and implemented to reduce the vulnerabilities. A vulnerability assessment can be a standalone process or part of a full risk assessment. During this assessment, the analysis of site assets is based on a) the identified threat, b) the criticality of the assets, and c) the level of protection chosen (i.e., based on willingness or unwillingness to accept risk).

1. Much of the information in this section is from U.S. Department of Energy (DOE 2002), *Vulnerability Assessment Methodology: Electric Power Infrastructure*; U.S. Army Research Laboratory (2000), *Vulnerability Risk Assessment, ARL-TR-1045*; U.S. Dept of Justice (2002), *A Method to Assess the Vulnerability of U.S. Chemical Facilities.*

It is important to point out that post 9/11 all sectors have taken great strides to protect their critical infrastructure. For instance, government and industry have developed vulnerability assessment methodologies for critical infrastructure systems and trained thousands of auditors and others to conduct them.

The actual complexity of vulnerability assessments will range based upon the design and operation of the DIB asset. The nature and extent of the VA will differ among systems based on a number of factors, including system size, potential population, and safety evaluations, also vary based on knowledge and types of threats, available security technologies, and applicable local, state, and federal regulations. Preferably, a VA is "performance-based," meaning that it evaluates the risk to the DIB assets based on the effectiveness (performance) of existing and planned measures to counteract adversarial actions. According to USEPA (2002), the common elements of DIB sector vulnerability assessments are as follows:

- Characterization of the DIB sector, including its mission and objectives
- Identification and prioritization of adverse consequences to avoid
- Determination of critical assets that might be subject to malevolent acts that could result in undesired consequences
- Assessment of the likelihood (qualitative probability) of such malevolent acts from adversaries
- Evaluation of existing countermeasures
- Analysis of current risk and development of a prioritized plan for risk reduction

DIB sector members should routinely perform vulnerability assessments to better understand threats and vulnerabilities, determine acceptable levels of risk, and stimulate action to mitigate identified vulnerabilities. The direct benefits of performing a vulnerability assessment include:

- **Build and broaden awareness**—the assessment process directs senior management's attention to security. Security issues, risks, vulnerabilities, mitigation options, and best practices are brought to the surface. Awareness is one of the least expensive and most effective methods for improving the organization's overall security posture.

- **Establish or evaluate against a baseline**—if a baseline has been previously established, an assessment is an opportunity for a checkup to gauge the improvement or deterioration of security. If no previous baseline has been performed (or the work was not uniform or comprehensive), an assessment is an opportunity to integrate and unify previous efforts, define common metrics, and establish a definitive baseline. The baseline also can be compared against best practices to provide perspective on an organization's security posture.
- **Identify vulnerabilities and develop responses**—generating lists of vulnerabilities and potential responses is usually a core activity and outcome of an assessment. Sometimes, due to budget, time, complexity, and risk considerations, the response selected for many of the vulnerabilities may be non-action, but after completing the assessment process these decisions will be conscious ones, with a documented decision process and item-by-item rationale available for revisiting issues at scheduled intervals. This information can help drive or motivate the development of a risk management process.
- **Categorize key assets and drive the risk management process**—an assessment can be a vehicle for reaching enterprise-wide consensus on a hierarchy of key assets. This ranking, combined with threat, vulnerability, and risk analysis, is at the heart of any risk management process. For many organizations, "Y2K" that was the first time a company-wide inventory and ranking of key assets was attempted. An assessment allows any organization to revisit that list form a broader and more comprehensive perspective.
- **Develop and build internal skills and expertise**—a security assessment, when not implemented in an "audit" mode, can serve as an excellent opportunity to build security skills and expertise within an organization. A well-structured assessment can have elements that serve as a forum for cross-cutting groups to come together and share issues, experiences, and expertise. External assessors can be instructed to emphasize "teaching and collaborating" rather than "evaluating" (the traditional role). Whatever the organization's current level of sophistication, a long-term goal should be to move the organization toward a capability for self-assessment.
- **Promote action**—although disparate security efforts may be underway in an organization, an assessment can crystallize and focus attention and resources on solving specific and systemic security problems. Often the people in the

trenches are well aware of security issues (and even potential solutions) but are unable to convert their awareness to action. An assessment provides an outlet for their concerns and the potential to bring these issues to the appropriate levels (legal, financial, executive) and achieve action.

- **Kick off an ongoing security effort**—an assessment can be used as a catalyst to involve people throughout the organization in security issues, build cross-cutting teams, establish permanent forums and councils, and harness the momentum generated by the assessment to build an ongoing institutional security effort. The assessment can lead to the creation of either an actual or a virtual (matrixed) security organization.

THE VULNERABILITY ASSESSMENT PROCESS

Table 5.1 provides an overview of the elements included in the assessment methodology. The elements included in this overview are based on actual in-field experience and lessons learned.

In Table 5.1, step 3 deals with identification of asset criticality. This is an important step in any vulnerability assessment. Identifying asset criticality serves several functions:

- It enables more careful consideration of factors that affect risk, including threats, vulnerabilities, and consequences of loss or compromise of the asset.
- It enables more focused and thorough consideration of loss or compromise of the asset.
- It enables leaders to develop robust methods for managing consequences of asset loss (restoration).
- It provides a means to increase awareness of a broad range of employees to protect truly critical assets and to differentiate in policies and procedures the heightened protection they require.

As previously indicated, identifying the criticality of assets is used primarily to focus the vulnerability analysis efforts. It also assists with the ranking of various recommendations for reducing vulnerabilities. Let's take a look at the criticality of electric power assets and operations included in the normal operation of DIB sector assets.

Table 5.1. Basic Elements in Vulnerability Assessments

Element	*Points to Consider*
1. Characterization of the DIB entity, including its mission and objectives	• What are the important missions of the system to be assessed? Define the highest priority services provided by the utility. Identify the industry's customers: * General public * Government * Military * Industrial * Critical care * Retail operations * Firefighting • What are the most important facilities, processes, and assets of the system for achieving the mission objectives and avoiding undesired consequences? Describe the: * Industry facilities * Operating procedures * Management practices that are necessary to achieve the mission objectives * How the industry operates * Treatment processes * Storage methods and capacity * Energy use and storage * Distribution system In assessing those assets that are critical, consider critical customers, dependence on other infrastructures (e.g., chemical, transportation, communications), contractual obligations, single points of failure, chemical hazards and other aspects of the industry's operations, or availability of industry utilities that may increase or decrease the criticality of specific facilities, processes and assets.
2. Identification and prioritization of adverse consequences to avoid	• Take into account the impacts that could substantially disrupt the ability of the system to provide a safe and reliable supply of materials. DIB sector systems should use the vulnerability assessment process to determine how to reduce risk associated with the consequences of significant concern. • Ranges of consequences or impacts for each of these events should be identified and defined. Factors to be considered in assessing the consequences may include:

Element	Points to Consider
	* Magnitude of service disruption * Economic impact (such as replacement and installation costs for damaged critical assets or loss of revenue due to service outage) * Number of illnesses or deaths resulting from an event * Impact on public confidence in the material supply * Chronic problems arising from specific events * Other indicators of the impact of each event as determined by the DIB sector Risk reduction recommendations at the conclusion of the vulnerability assessment strive to prevent or reduce each of these consequences.
3. Determination of critical assets that might be subject to malevolent acts	• What are the malevolent acts that could reasonably cause undesired? * Electronic, computer, or other automated systems which are utilized by the DIB sector entities (e.g., Supervisory Control and Data Acquisition [SCADA]) * The use, storage, or handling of various DIB materials (oil, coal, gas, etc.) * The operation and maintenance of such systems
4. Assessment of the likelihood (qualitative probability) of such malevolent acts from adversaries (e.g., terrorists, vandals)	• Determine the possible modes of attack that might result in consequences of significant concern based on critical assets of the DIB sector entity. The objective of this step of the assessment is to move beyond what is merely possible and determine the likelihood of a particular attack scenario. This is a very difficult task as there is often insufficient information to determine the likelihood of a particular event with any degree of certainty. • The threats (the kind of adversary and the mode of attack) selected for consideration during a vulnerability assessment will dictate, to a great extent, the risk reduction measures that should be designed to counter the threat(s). Some vulnerability assessment methodologies refer to this as a "Design Basis Threat" (DBT) where the threat serves as the basis for the design of countermeasures, as well as the benchmark against which vulnerabilities are assessed.

(*continued*)

Table 5.1. *Continued*

Element	*Points to Consider*
	Differences in geographic location, size of the utility, previous attacks in the local area, and many other factors will influence the threat(s) that the DIB sector entity should consider in their assessments. DIB sector entities should consult with the local FBI and/or other law enforcement agencies, public officials, and others to determine the threats upon which their risk reduction measures should be based.
5. Evaluation of existing countermeasures	• What capabilities does the system currently employ for detection, delay, and response?
(Depending on countermeasures already in place, some critical assets may already be sufficiently protected. This step will aid in identification of the areas of greatest concern and help to focus priorities for risk reduction.)	* Identify and evaluate current detection capabilities such as intrusion-detection systems, energy quality monitoring, operational alarms, guard post orders, and employee security awareness programs. * Identify current delay mechanisms such as locks and key control, fencing, structure integrity of critical assets, and vehicle access checkpoints. * Identify existing policies and procedures for evaluation and response to intrusion and system malfunction alarms, and cybersystem intrusions. **It is important to determine the performance characteristics. Poorly operated and maintained security technologies provide little or no protection.** • What cyber protection system features does the facility have in place? Assess what protective measures are in place for the SCADA and business-related computer information systems such as: * Firewalls * Modem access * Internet and other external connections, including wireless data and voice communications * Security polices and protocols **It is important to identify whether vendors have access rights and/or "backdoors" to conduct system diagnostics remotely.** • What security policies and procedures exist, and what is the compliance record for them? Identify existing policies and procedures concerning:

Element	Points to Consider
	* Personal security * Physical security * Key and access badge control * Control of system configuration and operational data * Chemical and other vendor deliveries * Security training and exercise records
6. Analysis of current risk and development of a prioritized plan for risk reduction	• Information gathered on threat, critical assets, DIB sector operations, consequences, and existing countermeasures should be analyzed to determine the current level of risk. The utility should then determine whether current risks are acceptable or risk reduction measures should be pursued. • Recommended actions should measurably reduce risks by reducing vulnerabilities and/or consequences through improved deterrence, delay, detection, and/or response capabilities or by improving operational policies or procedures. Selection of specific risk reduction actions should be completed prior to considering the cost of the recommended action(s). Facilities should carefully consider both short- and long-term solutions. An analysis of the cost of short- and long-term risk reduction actions may impact which actions the utility chooses to achieve its security goals. • Facilities may also want to consider security improvements. Security and general infrastructure may provide significant multiple benefits. For example, improved treatment processes or system redundancies can both reduce vulnerabilities and enhance day-to-day operation. • Generally, strategies for reducing vulnerabilities fall into three broad categories: * Sound business practices—affect policies, procedures, and training to improve the overall security-related culture at the facility * System upgrades—include changes in operations, equipment, processes, or infrastructure itself that make the system fundamentally safer * Security upgrades—improve capabilities for detection, delay, or response

Adapted from Spellman (2009), *The Handbook of Safety Engineering.*

Physical

- Generators
- Substations
- Transformers
- Transmission lines
- Distribution lines
- Control center
- Warehouses
- Office buildings
- Internal and external infrastructure dependencies

Cyber

- SCADA systems
- Networks
- Databases
- Business systems
- Telecommunications

Interdependencies

- Single-point nodes of failures
- Critical infrastructure components of high reliance

VULNERABILITY ASSESSMENT METHODOLOGY

Vulnerability assessment methodology consists of ten elements. Each element along with a fuller description is listed next (USDOE 2002).

1. Network architecture
2. Threat environment
3. Penetration testing
4. Physical security
5. Physical asset analysis
6. Operations security
7. Policies and procedures
8. Impact analysis
9. Infrastructure interdependencies
10. Risk characterization

Network Architecture

This element provides an analysis of the information assurance features of the information network(s) associated with the organization's critical information systems. Information examined should include network topology and connectivity (including subnets), principal information assets, interface and communication protocols, function and linage of major software and hardware components (especially those associated with information security such as intrusion detectors), and policies and procedures that govern security features of the network.

Procedures for information assurance in the system, including authentication of access and management of access authorization, should be reviewed. The assessment should identify any obvious concerns related to architectural vulnerabilities, as well as operating procedures. Existing security plans should be evaluated, and the results of any prior testing should be analyzed. Results from the network architecture assessment should include potential recommendations for changes in the information architecture, functional areas and categories where testing is needed, and suggestions regarding system design that would enable more effective information and information system protection.

Three techniques are often used in conduction the network architecture assessment:

1. Analysis of network and system documentation during and after the site visit
2. Interview with facility staff, managers, and chief information officer
3. Tours and physical inspections of key facilities

Threat Environment

Development of a clear understanding of the threat environment is a fundamental element of risk management. When combined with an appreciation of the value of the information assets and systems, and the impact of unauthorized access and subsequent malicious activity, an understanding of threats provides a basis for better defining the level of investment needed to prevent such access.

The threat of a terrorist attack to DIB sector infrastructure could come from several areas, including physical, cyber, and interdependency. In addition, threats could come from individuals or organizations motivated by financial gain or persons who derive pleasure from such penetration (e.g., recreational hackers, disgruntled employees). Other possible sources of threats are those who want to accomplish extremist goals (e.g., environmental terrorists, antinuclear advocates) or embarrass one or more organizations.

This element should include a characterization of these and other threats, identification of trends in these threats, and ways in which vulnerabilities are exploited. To the extent possible, characterization of the threat environment should be localized—that is, within the organization's service area.

Penetration Testing

The purpose of network penetration testing is to utilize active scanning and penetration tools to identity vulnerabilities that a determined adversary could easily exploit. Penetration testing can be customized to meet the specific needs and concerns of the DIB sector unit. In general, penetration testing should include a test plan and details on the rules of engagement (ROE). It should also include a general characterization of the access points of the critical information systems and communication interface connections, modem network connections, access points to principal network routers, and other external connections. Finally, penetration testing should include identified vulnerabilities and, in particular, whether access could be gained to the control network or specific subsystem or devices that have a critical role in assuring continuity of service.

Penetration testing consists of an overall process of establishing the ground rules or ROE for the test; establishing a white cell for continuous communication; developing a format or methodology for the test; conducting the test; and generating a final report that details methods, findings, and recommendations.

Penetration testing methodology consists of three phases: reconnaissance, scenario development, and exploitation. A one-time penetration test can provide the utility with valuable feedback; however, it is far more effective if performed on a regular basis. Repeated testing is recommended because new threats develop continuously, and the networks, computers, and architecture of the DIB sector unit or utility are likely to change over time.

Physical Security

The purpose of physical security assessment is to examine and evaluate the systems in place (or being planned) and to identify potential improvements in this area for the sites evaluated. Physical security systems include access controls, barriers locks and keys, badges and passes, intrusion-detection devices and associated alarm reporting and display, closed-circuit television (assessment and surveillance), communications equipment (telephone, two-way radio, intercom, cellular), lighting (interior and exterior), power sources (line, battery, generator), inventory control, postings (signs), security system wiring, and protective force. Physical security systems are reviewed for design, installation operation, maintenance, and testing.

The physical security assessment should focus on those sites directly related to the critical facilities, including information systems and assets required for operation. Typically include are facilities that house critical equipment or information assets or networks dedicated to the operation of electric, oil, or gas transmission, storage, or delivery systems. Other facilities can be included on the basis of criteria specified by the organization being assessed. Appropriate levels of physical security are contingent upon the value of company assets, the potential threats to these assets, and the cost associated with protecting the assets. Once the cost of implementing/maintaining physical security programs is known, it can be compared to the value of the company assets, thus providing the necessary information for risk management decisions. The focus of the physical security assessment task is determined by prioritizing the company assets; that is, the most critical assets receive the majority of the assessment activity.

At the start of the assessment, survey personnel should develop a prioritized listing of company assets. This list should be discussed with company personnel to identify areas of security strengths and weaknesses. During these initial interviews, assessment area that would provide the most benefit to the company should be identified; once known, they should become the major focus of the assessment activities.

The physical security assessment of each focus area usually consists of the following:

- Physical security program (general)
- Physical security program (planning)
- Barriers
- Access controls/badges
- Locks/keys

- Intrusion-detection systems
- Communications equipment
- Protective force/local law enforcement agency

The key to reviewing the above topics is not to just identify if they exist but to determine the appropriate level that is necessary and consistent with the value of the asset being protected. The physical security assessment worksheets provide guidance on appropriate levels of protection.

Once the focus and content of the assessment task have been identified, the approach to conduction the assessment can be either at the "implementation level" or at the "organizational level." The approach taken depends on the maturity of the security program.

For example, a company with a solid security infrastructure (staffing plans/procedures, funding) should receive a cursory review of these items; however, facilities where the security programs are being implemented should receive a detailed review. The security staff can act upon deficiencies found at the facilities, once reported.

For companies with an insufficient security organization, the majority of time spent on the assessment should take place at the organizational level to identify the appropriate staffing/funding necessary to implement security programs to protect company assets. Research into specific facility deficiencies should be limited to finding just enough examples to support any staffing/funding recommendations.

Physical Asset Analysis

The purpose of the physical asset analysis is to examine the systems and physical operational assets to ascertain whether vulnerabilities exist. Included in this element is an examination of asset utilization, system redundancies, and emergency operating procedures. Consideration should also be given to the topology and operating practices for electric and gas transmission, processing, storage, and delivery, looking specifically for those elements that either singly or in concert with other factors provides a high potential for disrupting service. This portion of the assessment determines company and industry trends regarding these physical assets. Historic trends, such as asset utilization, maintenance, new infrastructure investments, spare parts, SCADA linkages, and field personnel are part of the scoping element.

The proposed methodology for physical assets is based on a macro-level approach. The analysis can be performed with company data, public data, or both. Some companies might not have readily available data or might be reluctant to share that data.

Key output from analysis should be graphs that show trends. The historic data analysis should be supplemented with on-site interviews and visits. Items to focus on during a site visit include the following:

- Trends in field testing
- Trends in maintenance expenditures
- Trends in infrastructure investments
- Historic infrastructure outages
- Critical system components and potential system bottlenecks
- Overall system operation controls
- Use and dependency of SCADA systems
- Linkages of operation staff with physical and IT security
- Adequate policies and procedures
- Communications with other regional utilities
- Communications with external infrastructure providers
- Adequate organizational structure

Operations Security

Operations security (OPSEC) is the systematic process of denying potential adversaries (including competitors or their agents) information about capabilities and intentions of the host organization. OPSEC involves identifying, controlling, and protecting generally nonsensitive activities concerning planning and execution of sensitive activities. The OPSEC assessment reviews the processes and practices employed for denying adversary access to sensitive and nonsensitive information that might inappropriately aid or abet an individual's or organization's disproportionate influence over system operation. This assessment should include a review of security training and awareness programs, discussions with key staff, and tours of appropriate principal facilities. Information that might be available through public access should also be reviewed.

Policies and Procedures

The policies and procedures by which security is administered (1) provide the basis for identifying and resolving issues; (2) establish the standards of reference for policy implementation; and (3) define and communicate roles, responsibilities, authorities, and accountabilities for all individuals and organizations interfaced with critical systems. They are the backbone for decisions and day-to-day security operations. Security policies and procedures become particularly important at times when multiple parties must interact to effect a desired level of security and when substantial legal ramifications could result from policy violations. Policies and procedures should be reviewed to determine whether they (1) address the key factors affecting security; (2) enable effective compliance, implementation, and enforcement; (3) reference or conform to established standards; (4) provide clear and comprehensive guidance; and (5) effectively address roles, responsibilities, accountabilities, and authorities.

Understanding and assessing the policies and procedures area provide a means of identifying strengths and areas for improvements that can be achieved through:

- Modification of current policies and procedures
- Implementation of current policies and procedures
- Development and implementation of new policies and procedures
- Assurance of compliance with policies and procedures
- Cancellation of policies and procedures that are no longer relevant, or are inappropriate, for the facility's current strategy and operations

Impact Analysis

A detailed analysis should be conducted to determine the influence that exploitation of unauthorized access to critical facilities or information systems might have on an organization's operations (e.g., market and/or physical operations). In general, such an analysis would require thorough understanding of (1) the applications and their information processing, (2) decisions influenced by this information, (3) independent checks and balances that might exist regarding information upon which decisions are made, (4) factors that might mitigate the impact of unauthorized access, and (5) secondary impacts of such access. Similarly, the physical chain of events following disruption, including the primary, secondary, and tertiary impacts of disruption, should be examined.

The impact analysis provides an introduction to risk characterization by providing quantitative estimates of these impacts so that the DIB sector unit can implement a risk management program and weigh the risks and costs of various mitigation measures.

Infrastructure Interdependencies

The term "infrastructure interdependencies" refers to the physical and electronic (cyber) linkages within DIB and among our nation's critical infrastructures—energy (electric power, oil, natural gas), telecommunications, transportation, water supply systems, banking and finance, emergency services, and government services. This task identifies the direct infrastructure linkages between and among the infrastructures that support critical facilities as recognized by the organization. Performance of this task requires a detailed understanding of an organization's functions, internal infrastructures, and how these link to external infrastructures.

The purpose of the infrastructure interdependencies assessment is to examine and evaluate the infrastructures (internal and external) that support critical facility functions, along with their associated interdependencies and vulnerabilities.

Risk Characterization

Risk characterization provides a framework for prioritizing recommendations across all task areas. The recommendations for each task area are judged against a set of criteria to help prioritize the recommendations and assist the organization in determining the appropriate course of action. It provides a framework for assessing vulnerabilities, threats, and potential impacts (determined in the other tasks). In addition, the existing risk analysis and management process at the organization should be reviewed and, if appropriate, utilized for prioritizing recommendations. The degree to which corporate risk management includes security factors is also evaluated.

VULNERABILITY ASSESSMENT PROCEDURES

Vulnerability assessment procedures can be conducted for DIB sector assets using various methodologies. For example, the checklist analysis is an effective technology. In addition, Pareto analysis (80/20 principle), relative ranking, pre-removal risk assessment (PRRA), change analysis, failure mode and effects analysis (FMEA), fault tree analysis, event tree analysis, what-if analysis, and hazard and operability (HAZOP) can be used in conducting the assessment.

Based on personal experience, the what-if analysis and HAZOP seem to be the most user-friendly methodologies to use. A sample what-if analysis procedural outline is presented below, followed by a brief explanation and outline for conducting HAZOP.

What-If Analysis Procedure/Sample What-If Questions

The steps in a **What-If Checklist Analysis** are as follows:

1. Select the team (personnel experienced in the process).
2. Assemble information (piping and instrumentation drawings [P &IDs], process flow diagrams [PFDs], operating procedures, equipment drawings, etc.).
3. Develop a list of what-if questions.
4. Assemble your team in a room where each team member can view the information.
5. Ask each what-if question in turn and determine:
   - What can cause the deviation from design intent that is expressed by the question?
   - What adverse consequences might follow?
   - What are the existing design and procedural safeguards?
   - Are these safeguards adequate?
   - If these safeguards are not adequate, what additional safeguards does the team recommend?
6. As the discussion proceeds, record the answers to these questions in tabular format.
7. Do not restrict yourself to the list of questions that you developed before the project started. The team is free to ask additional questions at any time.
8. When you have finished the what-if questions, proceed to examine the checklist. The purpose of this checklist is to ensure that the team has not forgotten anything. While you are reviewing the checklist, other what-if questions may occur to you.
9. Make sure that you follow up all recommendations and action items that arise from the hazards evaluation.

HAZOP Analysis

The **HAZOP Analysis** technique uses a systematic process to (1) identify possible deviations from normal operations and (2) ensure that safeguards are in place to help prevent accidents. The HAZOP uses special adjectives (such as speed, flow, pressure, etc.) combined process conditions (such as "more," "less," or "no") to systematically consider all credible deviations from normal conditions. The adjectives, called guide words, are a unique feature of HAZOP analysis.

In this approach, each guide word (see table 5.2) is combined with relevant process parameters (see table 5.3) and applied at each point (study node, process section, or operating step) in the process that is being examined.

Table 5.2. HAZOP Analysis Guide Words

Guide Words	*Meaning*
No	Negation of the Design Intent
Less	Quantitative Decrease
More	Quantitative Increase
Part Of	Other Material Present by Intent
As Well As	Other Materials Present Unintentionally
Reverse	Logical Opposite of the Intent
Other Than	Complete Substitution

Table 5.3. Common HAZOP Analysis Process Parameters

Flow	*Time*	*Frequency*	*Mixing*
Pressure	Composition	Viscosity	Addition
Temperature	pH	Voltage	Separation
Level	Speed	Information	Reaction

The following is an example of creating deviations using guide words and process parameters:

Guide Words		**Parameter**		**Deviation**
NO	+	FLOW	=	NO FLOW
MORE	+	PRESSURE	=	HIGH PRESSURE
AS WELL AS	+	ONE PHASE	=	TWO PHASE
OTHER THAN	+	OPERATION	=	MAINTENANCE
MORE	+	LEVEL	=	HIGH LEVEL

Guide words are applied to both the more general parameters (e.g., react, mix) and the more specific parameters (e.g., pressure, temperature). With the general parameters, it is not unusual to have more than one deviation from the application of one guide word. For example, "more reaction" could mean either that a reaction takes place at a faster rate, or that a greater quantity of product results. On the other hand, some combination of guide words and parameters will yield no sensible deviation (e.g., "as well as" with "pressure").

HAZOP Procedure

1. Select the team.
2. Assemble information (P&IDs, PFDs, operating procedures, equipment drawings, etc.).
3. Assemble your team in a room where each team member can view P&IDs.
4. Divide the system you are reviewing into nodes (you can present the nodes, or the team can choose them as you go along).
5. Apply appropriate deviations to each node. For each deviation, address the following questions:
    - What can cause the deviation from design intent?
    - What adverse consequences might follow?
    - What are the existing design and procedural safeguards?
    - Are these safeguards adequate?
    - If these safeguards are not adequate, what does the team recommend?
6. As the discussion proceeds, record the answers to these questions in tabular format.

Checklist Procedure

One of the simplest vulnerability assessment methodologies to employ is the checklist. The Building Vulnerability Assessment Checklist developed by the Department of Veterans Affairs (VA) and is part of EEMA 426, *Reference Manual to Mitigate Potential Terrorist Attacks Against Buildings* is the reference manual that is recommended in this book and is available on the Internet at http://www.fema.gov/media-library-data/2013072. It is an excellent guide for conducting a viable checklist-type vulnerability assessment. This checklist will help you to prepare your threat assessment because it allows a consistent security evaluation of designs at various levels. The checklist can be used as screening tool for preliminary design vulnerability assessment and supports the preparation of all steps for use by the assessment teams during preparation for interviews with facility representatives to help assure that all relevant aspects of the DIB assets are considered in the survey.

REFERENCES AND RECOMMENDED READING

CBO (2004). *Homeland Security and the Private Sector*. Accessed 05/2/168 at www .cbo.gov/ft. docs.

CSI (2011). *2010/2011 Computer Crime and Security Survey*. Orlando, FL: Computer Security Institute.

DHS (2009). *National Infrastructure Protection Plan*. Retrieved 05/11/19 at http:// www.dhs.gov/ xlibrary/assets/NIPP.Plan.pdf.

DoD (2010). DoD Policy and Responsibility for Critical Infractions –DODD 3020.40. Washington, DC: U.S. Department of Defense.

FBI (2012). Insider Espionage. Report before the House Committee on Homeland Security Subcommittee on Counter Terrorism and Intelligence. Washington, DC: Federal Bureau of Investigation.

FEMA (2008). *FEMA452: Risk Assessment, A How-to Guide*. Accessed 05/01/16 at fema.gov/library/file?type=published/filetofile.

FEMA (2015). *Protecting Critical Infrastructure against Insider Threats*. Accessed 17 April 2015 at http://emilms.fema.gov/IS0915/IABsummary.htm.

National Infrastructure Advisory Council (2008). *First Report and Recommendations on the Insider Threat to Critical Infrastructure*. Washington, DC.

Spellman, F.R. (1997). *A Guide to Compliance for PSM/RMP*. Lancaster, PA: Technomic Publishing Company.

U.S. Army Research Laboratory (2000). *Vulnerability Risk Assessment, ARL-TR-1045*. Washington, DC: DOD

USDOE (2010). *Energy Sector-Specific Plan: An Annex to the National Infrastructure Protection Plan*. Washington, DC: USDOE.

USDOE (2002). *Vulnerability Assessment Methodology: Electric Power Infrastructure*. Washington, DC.

U.S. Dept of Justice (2002). *A Method to Assess the Vulnerability of U.S. Chemical Facilities*. Washington, DC.

USEPA (2002). *Vulnerability Assessment Fact Sheet*. EPA 816-F-02-025. Accessed May 2006 at www.epa.gov/ogwdw/security/index.html

6

Preparation

Question: When preparing to respond to terrorist acts against people, malls, schools, sports stadiums, libraries, and government facilities, when is enough, enough?

Answer: Until we can read the terrorists' minds, "enough" preparation is never enough. Simply, preparation is ongoing and never-ending.

—Frank R. Spellman

NEW YORK BOMBING: INVESTIGATORS SEARCH FOR SUSPECTS, MOTIVE

—CNN headline, 2016

We must take the battle to the enemy, disrupt his plans, and confront the worst threats before they emerge.

—George W. Bush

Now, obviously, the further removed we get from September 11, I think the natural tendency is to let down our guard. Unfortunately, we cannot do that. The government will continue to do everything we can to find and stop those who seek to harm us. And I believe we owe it to the American people to remind them that they must be vigilant, as well.

—Tom Ridge (quoted in Henry 2002)

The threat of a terrorist attack on the U.S. DIB infrastructure doesn't generate the same attention or fear as potential nuclear, biological, or chemical terrorism. It should. Yet when you ask anyone about the Defense Industrial Base, he or she just looks at you blankly. (I know, I *have* asked.) Still, because of the seriousness of the threat of terrorism to the nation's DIB sector and the enormous economic and security implications of such attacks, USDOE, USEPA, and other agencies have worked non-stop since 9/11 in gathering and providing as much advice and guidance as possible to aid DIB sector personnel in protecting DIB sector assets and associated critical support infrastructure. In this chapter, we provide an overview of important tools that can be used in protecting DIB sector to guard against the threat of terrorism.

DIB SECTOR THREATS AND INCIDENTS

Based on evidence of potential losses from past accidents, indication of the potential human and environmental losses and economic costs from an attack on a DIB sector facility or producer comes from major incidents that have occurred both abroad and in the United States. Those events indicate that the human and environmental losses could be significant (CBO 2004).

DIB sector threats and incidents may be of particular concern due to the range of potential consequences:

- Creating an adverse impact on public health within a population
- Disrupting system operations and interrupting the supply of critical military components
- Causing physical damage to system infrastructure
- Reducing public confidence in the DIB supply system
- Long-term denial of basic security and protection and the cost of replacement

Keep in mind that some of these consequences would be realized only in the event of a successful terrorist incident, but the mere threat of terrorism can also have an adverse impact on industries that depend on a safe, steady supply of DIB materials. In addition, the economic implications of such attacks are potentially enormous. For example, many believe that the reason we're looking at more than $60 for a barrel of oil ($61.68 at the beginning of 2018) is the fact that we have a "terror premium" factored in.

While it is important to consider the range of possibilities associated with a particular threat, assessments are typically based on the probability of a particular occurrence. Determining probability is somewhat subjective, and is often based on intelligence and previous incidents. As mentioned, there are historical accounts of accidental incidents that have caused tremendous death and destruction.

Threat Warning Signs[1]

A threat warning is an occurrence or discovery that indicates a potential threat that triggers an evaluation of the threat. It is important to note that these warnings must be evaluated in the context of typical industry activity and previous experience in order to avoid false alarms. Following is a brief description of potential warnings.

- **Security Breach**. Physical security breaches, such as unsecured doors, open hatches, and unlocked/forced gates, are probably the most common threat warnings. In most cases, the security breach is likely related to lax operations or typical criminal activity such as trespassing, vandalism, and theft. However, it may be prudent to assess *any* security breach with respect to the possibility of attack.
- **Witness Account**. Awareness of an incident may be triggered by a witness account of tampering. DIB sector sites/facilities should be aware that individuals observing suspicious behavior near DIB sector facilities/plants will likely call 911 and not the plant. In this case, the incident warning technically might come from law enforcement, as described below. Note: the witness may be a plant employee engaged in their normal duties.
- **Direct Notification by Perpetrator**. A threat may be made directly to the DIB sector site, plant, or facility, either verbally or in writing, including via Facebook and by Twitter. Historical incidents would indicate that verbal threats made over the phone are more likely than written threats. While the notification may be a hoax, threatening a DIB sector unit is a crime and should be taken seriously.

---

1. The following section is adapted from USEPA (2003), *Response Protocol Toolbox: Planning for and Responding to Drinking Water Contamination Threats and Incidents* (Washington, DC: United States Environmental Protection Agency).

- **Notification by Law Enforcement.** A DIB sector site/facility may receive notification about a threat directly from law enforcement, including local, county, state, or federal agencies. As discussed previously, such a threat could be a result of suspicious activity reported to law enforcement, either by a perpetrator, a witness, or the news media. Other information, gathered through intelligence or informants, could also lead law enforcement to conclude that there may be a threat to the DIB sector site/facility. While law enforcement will have to be lead in the criminal investigation, the DIB sector site/facility has primary responsibility for the safety of its equipment, energy-producing materials and processes, and public health. Thus, the plant's role will likely be to help law enforcement to appreciate the public health implications of a particular threat as well as the technical feasibility of carrying out a particular threat.
- **Notification by News Media.** A threat to destroy a DIB site/facility might be delivered to the news media, or the media may discover a threat. A conscientious journalist would immediately report such a threat to the police, and either the journalist or the police would immediately contact the DIB sector site/facility. This level of professionalism would provide an opportunity for the plant to work with the media and law enforcement to assess the credibility of the threat before any broader notification is made.

RESPONSE TO THREATS: WHEN IS ENOUGH, ENOUGH?

This section is *not* designed to discuss what specific steps to take in responding to a terrorist threat. Rather, the questions addressed in this section are: "Why is it necessary to plan to respond to DIB sector threats at all?" and "When have I done enough?"

Federal, state, and local programs already exist that—with varying degrees of effectiveness—encourage or require the operators of DIB sector sites/facilities to boost their efforts to promote safety and security and to share information that can help local governments plan for emergencies (Schierow 2004).

Proper planning is a delicate process because public health measures are rarely noticed or appreciated (like buried utility pipes, they are often hidden functions) except when they fail—then they are very visible. Defense Department consumers are particularly upset by unreliable DIB supply systems and DIB products (many of which are extremely potent and hazardous)—or practices that produce unsafe (contaminated) environmental media—water, air, and/or soil—because they are often viewed as entitlements, and indeed, it is reasonable for citizens to expect a high quality, safe environment. Public health failures during response to contamination threats often take the form of too much or too little action. The result of too little action, including no response at all, can have disastrous consequences that potentially result in public injuries or fatalities.

Considering the potential risks of an inappropriate response to a severe DIB hazardous material release or explosion threat, it is clear that a systematic approach is needed to evaluate threats. One overriding question is "When has a DIB producer done enough?" This question may be particularly difficult to address when considering the wide range of agencies that may be involved in a threat situation. Other organizations, such as USEPA, USDOE, CDC, USDOT, law enforcement agencies, public health departments, and so on, will each have unique obligations or interests in responding to a severe release or explosion threat.

The guiding principle for responding to threats is one of "due diligence" or "what is a suitable and sensible response to a threat?" As discussed above, some response to DIB hazardous material contamination threats is warranted due to the public health implications of an actual contamination incident. However, a DIB sector unit or munitions facility could spend a lot of time and money over-responding to every threat, which would be an ineffective use of resources.

Ultimately, the answer to the question of "due diligence" must be decided at the local level and will depend on a number of considerations. Among other factors, local authorities must decide what level of risk is reasonable in the context of a perceived threat. Careful planning is essential to developing an appropriate response to terrorist threats, and in fact, one primary objective of USEPA's *Response Protocol Toolbox* (RPTB) is to aid users in the development of their own site-specific plans that are consistent with the needs and responsibilities of the user. Beyond planning, the RPTB considers a careful evaluation of any terrorist threat, and an appropriate response based on the evaluation, to be the most important element of due diligence.

In the RPTB, the threat management process is considered in three successive stages: **possible**, **credible**, and **confirmed**. Thus, as the threat escalates through these three states, the actions that might be considered due diligence expand accordingly. The following paragraphs describe, in general terms, actions that might be considered as due diligence at these various stages.

- **Stage 1: "Is the threat possible?"** If a DIB facility is faced with a terrorism threat, they should evaluate the available information to determine whether or not the threat is "possible" (i.e., could something have actually happened). If the threat is possible, then immediate operational response actions might be implemented, and activities such as site characterization would be initiated to collect additional information to support the next stage of the threat evaluation.
- **Stage 2: "Is the threat credible?"** Once a threat is considered "possible," additional information will be necessary to determine if the threat is "credible." The threshold at the credible stage is higher than that at the possible stage, and in general there must be information to corroborate the threat in order for it to be considered credible.
- **Stage 3: "Has the incident been confirmed?"** Confirmation implies that definitive evidence and information have been collected to establish the presence of a threat. Obviously, at this stage the concept of due diligence takes on a whole new meaning since authorities are now faced with death and destruction and a potential public health crisis. Response actions at this point include all steps necessary to protect public health, property, and the environment.

PREPARATION

As an environmental/occupational safety, security and health professional consultant for various utilities and others on the East Coast, I have performed numerous pre-OSHA audit inspections and audits of various plant or facility PSM/RMP compliance programs. During these site visits, one factor always seemed to be universal. While conducting the plant/facility walk-around to gauge the organization's overall profile and status with OSHA and USEPA compliance, I almost always find that the plant/facility manager or superintendent who accompanied us was shocked to find out what was actually going on or taking place in their facilities. They would scratch their heads and ask various workers: "What the hell are you doing? Where did that new machine come from? When was it installed? Why is that door broken? Who told you to paint that door, machine or other apparatus? When did that hole get in the fence? Who left the back gate open? Where is the foreman" . . . and so on. Eventually, in an expression of utter consternation, the manager/superintendent asks, "Who the hell is in charge around here?" And this is basically the question I find myself asking—far too often.

In one inspection performed at a plant/facility right after 9/11, I drove up to the entrance gate and was impressed with the height and condition of the barbed-razor-wire-topped fence and gates. I could not enter through the gates until I identified myself over a speaker system while a CCTV camera focused on my face. I was let in and given instructions to sign in at the main office. Not bad, just the way it should be . . . or so I thought at the time. After walking most of the plant site in the company of the plant manager, we approached the back fence area which was close to a huge chlorine storage building. At the terminal end of the plant and fence, I noticed a large gate that was propped open with ivy growing on it, through it, and around it—obviously, the gate had been in the open position for quite some time. I asked the plant manager why the gate was open. He stated that it was always open . . . that it led to a downhill path to a beach area below where plant personnel had constructed a picnic area, fronting the James River.

We walked the path to the bottom picnic area, looked around, and then looked back up the path toward the open gate and the prominent structure standing within the chlorine storage building. While walking back up the path to the gate, I asked the plant manager if he was not concerned about the security and safety of the plant, because the gate was left open, and especially about the safety of the 50 tons of deadly chlorine gas stored in the chlorine storage building.

"Nah . . . no way . . . we are safe here. I really don't see anyone swimming up river just to get into the plant site, ha. Besides, we are surrounded by woods out here . . . there's nothing to attack anyway."

Once inside the plant, I asked the plant manager if he was not worried about terrorists or some disgruntled former employee using a boat filled with explosives or some other weapon(s) gaining easy access to the plant and especially the chlorine building via the James River beach landing and picnic area?

"Nah, that will never happen . . . who would be that stupid? There's nothing around here worth blowin' up!"

Later, when I checked the GIS/GPS system data and maps pertaining to and showing the plant and surrounding area, I noted that about one-half mile from the plant site was a large housing area, a brewery (with 700+ employees), and a very large theme and historical park—annually visited by more than 2,500,000 people each summer.

Know Your DIB Sector Production Systems

All DIB plant/facility managers and process equipment operating personnel must know their plants/facilities. For these persons, there is no excuse for not knowing every square inch of the plant/facility site. In particular, plant workers should know about any and all construction activities underway on the site; the actual construction parameters of the plant; and especially operation of all plant unit processes. In addition, plant management must not only know their operating staff but also their customers.

Construction and Operation

Each DIB sector plant/facility is unique with respect to age, operation, and complexity. Another aspect to be considered relates to the material-producing/using and manufacturing processing system. This is important, particularly in evaluating the potential spread of a spilled hazardous material or other potential contaminant. Propagation of production-chemical contaminants through a system is dependent on a number of factors, including mixing conditions at the point of contamination, hydraulic conditions within water systems at the time of the contaminant introduction, and reactions between the contaminant and other materials in the system or environment.

Information about construction materials used in the system may be contained within the plant records and can be useful in evaluating the fate and transport of a particular chemical contaminant through a system. For example, a particular contaminant may adsorb to the pipe material used in a utility's distribution system, and this type of information would be critical in evaluating remediation options following a chemical contamination incident.

Personnel

DIB sector employees are generally its most valuable asset in preparing for and responding to chemical contamination threats and incidents. They have knowledge of the system and hazards associated with the production chemicals. The importance of knowledgeable and experienced personnel is highlighted by the complexity of most chemical processing systems. This complexity makes a successful contamination of a specific target contingent upon detailed knowledge of the system configuration and usage patterns. If perpetrators have somehow gained a sophisticated understanding of a chemical production process, the day-to-day experience of material production or chemical system personnel will prove an invaluable tool to countering any attacks. For instance, personnel may continually look for unusual aspects of daily operation that might be interpreted as a potential threat warning, and may also be aware of specific characteristics of the system that make it vulnerable to petrochemical contamination or worse.

Customers

A customer's knowledge of DIB supplies and chemicals is an important component of preventing and managing fires, explosions, and/or contamination incidents. Prevention is based largely on understanding potential targets of chemicals. Of special concern may be government buildings and military installations, or other institutions where large numbers of people could be directly or indirectly affected by a hazardous product/chemical threat or incident. Steps taken to protect the customer's employees and property, such as enhancements to the physical security of the customer's property these locations may deter the attack itself.

DIB customers vary significantly with regard to their expectations of what constitutes acceptable products and service, so it is necessary to consider the manner in which DIB products are used in a particular system. For example, high chemical demand that is largely driven by the defense industry has different implications compared to high usage rates in an urban center with a high population density. Some customers, such as factories and product manufacturers, may have certain chemical quality requirements. Sensitive subpopulations, including children and the elderly, can exhibit adverse health effects at doses more than an order of magnitude lower than those necessary to produce serious injury or death in a healthy adult. That being said, for the purposes of managing chemical fire, explosion, and contamination threats, it is important to keep in mind that the most important goal is protecting the health of the public as a whole. Planning, preparation, and allocation of resources should be directed toward protecting the public at large, beyond specific demographic groups or individual users.

Update ERP for Intentional Spill, Explosion, or Contamination

ERPs (emergency response plans) are nothing new to DIB sector facilities/plants, since many have developed ERPs to deal with natural disasters, accidents, violence in the workplace, civil unrest, and so on. Because the DIB sector and its related industries are a vital ingredient in protecting and sustaining our way of life, it has been prudent (and mandated) for DIB sector entities to develop ERPs in order to help ensure the continuous flow of defense industry products to their customers. However many DIB sector ERPs developed prior to 9/11 do not explicitly deal with terrorist threats, such as intentional fire, explosion, or contamination. Recently, the U.S. Congress and federal regulators have required DIB sector sites and especially hazardous chemical/products industries to prepare or revise, as necessary, an ERP to reflect the findings of their vulnerability assessment and to address terrorist threats (DHS 2010).

Establish Communication and Notification Strategy

Communication strategies must be planned and made available to all potential participants prior to an actual incident or threat. For the purposes of responding to a DIB sector threat, the communication structure could have several management levels within the industry, as well as external to the industry (governmental emergency response units, for example), that may be involved in management of a threat. The hierarchy of potential participants includes: the DIB sector site/facility, local government, regional government (e.g., county), state government, and federal government. Not all of these levels would necessarily be involved in every situation; however, the mechanism and process through which they interact must be decided in advance of an incident to achieve optimal public health and environmental protection. For any type of DIB materials incident, CHEMTREC (CHEMical Transportation Emergency Center) is dedicated to assisting emergency responders deal with incidents involving a hazardous material is available 24 hours each day. Due to the number and variety of possible participants, planning for effective communication is critical.

Perform Training and Desk/Field Exercises

In addition to a lack of planning, another reason that emergency response plans fail is lack of training and practice. Training provides the necessary means for everyone involved to acquire the skills to fulfill their role during an emergency. It may also provide important "buy-in" to the response process from both management and staff, which is essential to the success of any response plan. Desk exercises (also known as "tabletops" or "dry runs") along with field exercises allow participants to practice their skills. Also, these exercises will provide a test of the plan itself, revealing strengths and weakness that may be used to improve the overall plan. Improvements can include measures not only for intentional explosion of explosives, fuels, or petrochemicals or contamination of an environment with these materials, but also for other emergencies faced by the DIB sector and the community at large.

Enhance Physical Security

Where possible deny physical access to key sites; within the DIB sector system this may act as a deterrent to a perpetrator. When we consider that many of the DIB sector units such as munitions manufacturers and distribution centers are often in remote, wide-open spaces, this can be a huge challenge. Terrorists often seek the easiest route of attack, just like a burglar prefers a house with an open window or an open automobile with keys in the ignition. Aside from deterring actual attacks, enhancing physical security has other benefits. For example, installation of fences and locks may reduce the rate of false alarms. Without surveillance equipment or locks, it may not be possible to determine whether a suspicious individual has actually entered a vulnerable area. The presence of a lock and a determination as to whether it has been cut or broken provides sound, although not definitive, evidence that an intrusion has occurred. Likewise, security cameras can be used to review security breaches and determine if the incident was simply due to trespassing or is a potential contamination threat. The costs of enhancing physical security may be justified by comparison to the cost of responding to just one credible munitions explosion or contamination threat involving site characterization and lab analysis for potential contaminants.

Establish Baseline Monitoring Program

Background concentrations of suspected or tentatively identified hazardous chemical contaminants may be extremely important in determining if an incident has occurred. In some cases, and for some hazardous chemicals, background levels may be at detectable concentrations. Baseline occurrence information is derived from monitoring data and is used to characterize typical levels of a particular chemical contaminant.

SITE CHARACTERIZATION AND SAMPLING

Site characterization is defined as the process of collecting information from an investigation site in order to support the evaluation of a hazardous chemical-munitions-based incident that could lead to a fire, explosion, or contamination threat. Site characterization promotes and recommends the scientific method of inquiry using the conceptual site modeling process to characterize the environmental condition of a site using a multi-phased approach. Site characterization activities include the site evaluation, field safety screening, rapid field testing of environmental media (air, water, soil), and sample collection. The investigation site is the focus of site characterization activities, and if a suspected contamination site has been identified, it will likely be designated as the primary investigation site. Additional or secondary investigation sites may be identified due to the potential spread of a suspected contaminant. The results of site characterization are of critical importance to the threat evaluation process.

There are two broad phases of site characterization: planning and implementation. The incident commander is responsible for planning, while the site characterization team is responsible for implementing the site characterization plan. This section is intended as resource for those involved in either the planning or implementation phases of site characterization. While the target audience is primarily DIB industry managers and staff, other organizations (e.g., police, fire departments, FBI, and USEPA criminal investigators) may be involved in site characterization activities.

Site Characterization Process

USEPA's (2004; 2011) *Response Protocol Toolboxes* for water and wastewater contamination events are the models used here (because of their applicability to any critical infrastructures) to describe the following five stage DIB sector site characterization process:

1. **Customizing the Site Characterization Plan**. A site characterization plan is developed for a specific threat (possibly from a generic site characterization plan) and guides the team during site characterization activities.
2. **Approaching the Site**. Before entering the site, an initial assessment of site conditions and potential hazards is conducted at the site perimeter.

3. **Characterizing the Site**. The customized site characterization plan is implemented by conducting a detailed site investigation and rapid testing of the water, air, and soil.
4. **Collecting Samples**. Water/air/soil samples are collected in the event that lab analysis is required.
5. **Exiting the Site**. Following completion of site characterization, the site is secured and personnel exit the site and undergo any necessary decontamination.

While site characterization can be considered and implemented as a discrete process, it is important to regard it as an element of the threat evaluation process (see figure 6.1). In particular, site characterization is an activity initiated in response to a "possible" accidental or intentional fire, explosion, or contamination threat in order to gather information to help determine whether or not the threat is "credible." Initially, information from the threat evaluation supports the development of the customized site characterization plan. As this plan is implemented, the observations and results from site characterization feed into the threat evaluation. In turn, the revised threat evaluation may indicate that the threat is "credible," "not credible," or that the site characterization plan needs to be received in the field to collect more information in order to make this determination. Because threat evaluation and site characterization are interdependent, the incident commander must be in constant communication with the site characterization team while they are performing their tasks.

The first step is to develop a customized site characterization plan, which is based on the specific circumstances of the threat warning. This customized plan may be adapted from a generic site characterization plan, which is developed as part of a utility's preparation for responding to hazardous petrochemical threat. The site characterization team will use the customized plan as the basis for their activities at the investigation site. After an initial evaluation of available information, it is necessary to identify an investigation site where site characterization activities will be conducted. During the development of the customized plan, it is important to conduct an initial assessment of site hazards, which is critical to the safety of the site characterization team and may impact the makeup of the team. If there are obvious signs of hazards at the site, then teams trained in hazardous materials safety and handling

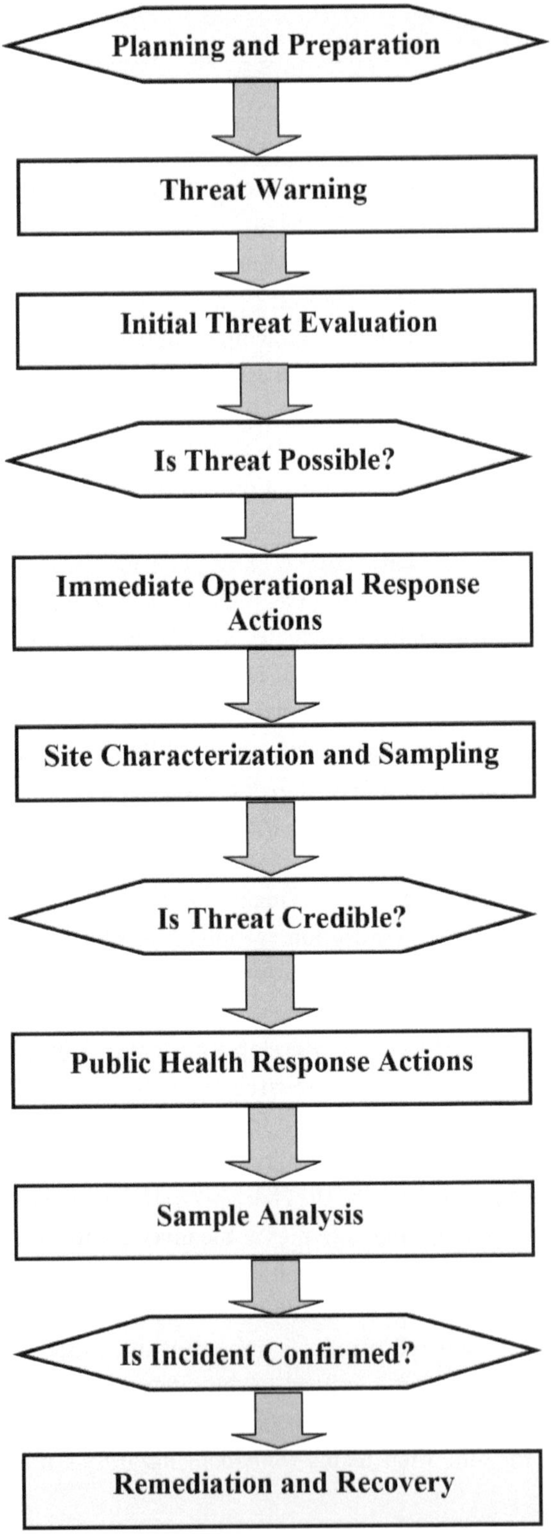

FIGURE 6.1
Threat Evaluation Process
USEPA (2004, 2011).

techniques may need to conduct an initial hazard assessment at the site and either "clear" the site for entry by utility personnel, or the HazMat team may decide to perform all site characterization activities themselves. Obvious signs of hazards would provide a basis for determining that a threat is "credible." Furthermore, the site might be considered a crime scene if there are obvious signs of hazards, and law enforcement may take over the site investigation.

Upon arrival at the site perimeter, the team first conducts field safety screening and observes site conditions. The purpose of field safety screening activities is to identify potential environmental hazards that might pose a risk to the site characterization team. The specific field safety screening performed should be identified in the site characterization plan, and might include screens for radioactivity and volatile organic chemicals (VOCs). If the team detects signs of hazard, they should stop their investigation and immediately contact the incident commander to report their findings.

If no immediate hazards are identified during approach to the site, the incident commander will likely approve the team to enter the site and perform the site characterization. During this stage, the team will continue field safety screening at the site, conduct a detailed site investigation, and perform rapid field testing of the water that is suspected of being contaminated.

Rapid field testing has three objectives: 1) provide additional information to support the threat evaluation process; 2) provide tentative identification of contaminants that would need to be confirmed later by lab testing; and 3) determine if hazards tentatively identified in the petrochemical release require special handling precautions. The specific rapid field testing performed should be identified in the site characterization plan. Specific field testing performed should be based on the circumstances of the specific threat and should consider the training, experience, and resources of the site characterization team. Negative field test results are not a reason to forgo petrochemical sampling, since field testing is limited in scope and can result in false negatives.

Following rapid field testing, samples of the potential contaminated air/water/soil will be collected for potential lab analysis. The decision to send samples to a lab for analysis should be based on the outcome of the threat evaluation. If the threat is determined to be "credible," then samples should be immediately delivered to the lab for analysis. The analytical approach for samples collected from the site should be developed with input from the supporting lab(s), based on information from the site characterization and threat evaluation. On the other hand, if the threat is determined to be "not credible," then samples should be secured and stored for a predetermined period in the event that it becomes necessary to analyze the samples at a later time.

Upon completion of site characterization activities, the team should prepare to exit the site. At this stage, the team should make sure that they have documented their findings, collect all equipment and samples, and re-secure the site (e.g., lock doors, hatches, and gates). If the site is considered to be a potentially hazardous site or crime scene, there may be additional steps involved in exiting this site.

Roles and Responsibilities

The incident commander and the site characterization team leader are key personnel in site characterization. The incident commander has overall responsibility for managing the response to the threat, and is responsible for planning and directing site characterization activities. The incident commander may also approve the site characterization team to proceed with their activities at key decision points in the process (e.g., whether or not to enter the site following he approach).

The site characterization team leader is responsible for implementing the site characterization plan in the field and supervising site characterization personnel. The site characterization team leader must coordinate and communicate with the incident commander during site characterization.

Depending on the nature of the contamination treat, other agencies and organizations may be involved or otherwise assume some responsibility during planning and implementation of site characterization activities. However, it is the incident commander has the ultimate responsibility for determining the scope of the site characterization activities and the team makeup.

Planning for Site Characterization

Providing training of staff involved in site characterization and sampling activities is critical. Responding to the site of a potential contamination incident is very different from routine inspection and sampling activities performed by utility staff. The equipment and safety procedures used at the site of a potential contamination incident may differ significantly from those used during more typical field activities. Providing staff training in the procedures presented in this section will help to ensure that they are properly and safely implemented during emergency situations.

Safety Practices and Personnel Protection

Proper safety practices are essential for minimizing risk to the DIB sector site characterization team and must be established prior to an incident in order to be effective. Field personnel involved in site characterization activities should have appropriate safety training to conform to appropriate regulations, such as OSHA 1910.120 and U.S. DoD directives, which deal with hazardous and deadly substances and materials. If planners and field personnel do not conclude that these regulations are applicable to them, they may still wish to adopt some of the safety principles in these regulations. The following guidance is provided to help the user develop his or her safety policies and practices. These safety policies should be consistent with the equipment and capabilities of the site characterization team and any applicable regulations.

The appropriate level of personal protection necessary to safely perform the site characterization activities will depend on the assessment of site hazards that might pose a risk to the site characterization team. The hazard assessment may be further refined during the approach to the site, based on the results of the field safety screening and initial observations of site conditions. Two general scenarios are considered, one in which there are no obvious signs of immediate hazards, and one in which there are indicators of site hazards.

Sample Collection Kits and Field Test Kits

Two types of kits are discussed in this section, sample collection kits and field test kits. Sample collection kits will generally contain all sample containers, materials, supplies, and forms necessary to perform sample collection activities. Field test kits contain the equipment and supplies necessary to perform field safety screening and rapid field testing of the air, water, and/or soil. Sample collection kits will generally be less expensive to construct than field test kits, and by constructing these two types of kits separately, sample collection kits can be pre-positioned throughout a system while the more expensive field kits may be assigned to specific site characterization teams or personnel.

The design and construction of sample collection and field test kits is a planning activity, since these kits must be ready to go at a moment's notice in response to a "possible" contamination threat. In addition to improving the efficacy of the site characterization and sampling activities, advance preparation of sample collection and field test kits offers several advantages:

- Sample collection and field test kits can be standardized throughout an area to facilitate sharing of kits in the event of an emergency that requires extensive sampling.
- Collection of a complete sample set is more likely to be achieved through the use of predesigned kits.
- Sample collection kits can be pre-positioned at key locations to expedite the sampling process.
- Personnel responsible for site characterization can become familiar with the content of the kits and trained in the use of any specialized equipment.

Generic Site Characterization Plan

A site characterization plan is developed to provide direction and communication between the incident commander and the site characterization team, which will facilitate the safe and efficient implementation of site characterization activities. The plan should be developed expeditiously since the site characterization results are an important input to the threat evaluation process. The rapid development of a site characterization plan can be facilitated by the development of a generic site characterization plan, which is easily customized to a specific situation. While the circumstances of a particular threat warning will dictate the specifics of a customized site characterization plan, many activities and procedures will remain the same for most situations, and these common aspects can be documented in the generic site characterization plan. Potential element of a generic plan include: pre-entry criteria, communications, team organization and responsibilities, safety, field testing, sampling, and exiting the site.

Pre-entry criteria define the conditions and circumstances under which site characterization activities will be initiated and the manner in which these activities will proceed. At each stage of the process (i.e., approach to the site, on-site characterization activities, sample collection, and exiting the site), specific criteria may be defined for proceeding to the next stage. The pre-entry criteria may also specify the general makeup of the site characterization team under various circumstances. For example, under low hazard conditions petrochemical facility teams may perform site characterization, while specially trained responders might be called upon to assist in the case of potentially hazardous conditions at the site. The criteria developed for a particular petrochemical facility should be consistent with the role that the facility has assumed in performing site characterization activities

The generic plan should define communication processes to ensure rapid transmittal of findings and a procedure for obtaining approval to proceed to the next stage of site characterization. It is advisable for the site characterization team to remain in constant communication with the incident commander for the entire time that they are on site. The plan should provide an approval process for the team to advance through the approach and on-site evaluation stages of the characterization, to ensure that the team is not advancing into a hazardous situation. Communication devices (e.g., cell phone, two-way radio, or panic button) can be used to alert incident command of problems/observations encountered in the field. The communication section of the generic plan should also discuss coordination with other agencies (e.g., law enforcement, fire department) and contingencies for contacting HazMat responders.

Field testing and sampling may be handled in the generic plan by presenting a menu that covers all potential option available to the utility, based on both internal and external capabilities. In developing a customized plant, the incident commander can simply check off the field tests and sampling requirements that are appropriate for the specific situation. The site characterization plan may also need to be revised in the field based on the observations of the team.

Quality Assurance for Field Testing and Sampling

Because of the diversity of potential field testing and sampling activities during the characterization, there may be no specific quality assurance (QA) activities which apply to all sampling procedures. However, the following general QA principles would apply in most cases and are consistent with the QA guidelines published by USEPA's Environmental Response Team:

- All data should be documented on field data sheets or within site logbooks.
- All instrumentation should be operated in accordance with operating instructions as supplied by the manufacturer, unless otherwise specified in the work plan. Equipment checkout and calibration activities should occur prior to site characterization and documented.

- Any relevant QA principles and plans specific to the particular water utility or responding organization should be observed.

Maintaining Crime Scene Integrity

When a DIB asset is attacked, destroyed, and/or contamination is released from the site, the focus of site characterization activities could potentially become the scene of a criminal investigation. If law enforcement takes responsibility for incident command because they believe a crime has been committed, they will control the site and dictate how any additional activities, such as site characterization, are performed. In cases in which the DIB facility is still responsible for incident command, it may still be prudent to take precautions to maintain the integrity of the potential crime scene during site characterization activities. The following guidelines for maintaining crime scene integrity are provided, although this should not necessarily be considered an exhaustive list:

- If there is substantial physical evidence of contamination at a site, the threat will likely be deemed "credible" from a utility and a law enforcement perspective. In this case, law enforcement may take control of the site and limit the activities performed by other organizations at the site.
- Substantial physical evidence of contamination might include discarded PPE, equipment (such as pumps and hoses), or containers with residual material. Special care should be taken to avoid moving or disturbing any potential physical evidence.
- Evidence should not be handled except at the direction of the appropriate law enforcement agency. Specially trained teams from the law enforcement community are best suited (and may be jurisdictionally required) for the collection of physical evidence from a contaminated crime scene.
- The collection of physical evidence is not generally considered time sensitive; however, site characterization and sampling activities are time sensitive due to the public health implications of contaminated environmental media: air, water, and/or soil. Thus, collection of environmental media samples may precede collection of physical evidence, and care must be taken not to disturb the crime scene while performing these activities. If samples can be collected outside of the boundaries of the suspected crime scene, it may avoid concerns about the integrity of the crime scene.

- Water, air, and/or soil samples collected for the purpose of confirming/dismissing a contamination threat and identifying a contaminant could potentially be considered evidence and should be handled accordingly.
- Because the analytical results may be considered evidence as well it is important to use a qualified lab for analytical support. If law enforcement has taken control of the situation prior to sample collection, they may require the collection of an additional sample set to be analyzed by their designated lab.
- Photographs and videos can be taken during the site characterization for use in the criminal investigation. Law enforcement should be consulted for proper handling during and after taking photographs/videos to ensure integrity of the evidence.

Maintaining crime scene integrity during site characterization is largely an awareness issue. If the site characterization team integrates the guidelines outline above into their on-site activities, they will go a long way toward maintaining the integrity of the crime scene.

Site Characterization Report

In order to provide useful information to support the threat evaluation process and the development of an analytical approach, the findings of the site characterization should be summarized in a report. This report is not intended to be a formal document, but simply a concise summary of information from the site activities that can be quickly assembled within an hour or two. The recommended content of the report includes:

- General information about the site
- Information about potential site hazards (keep in mind that many hazards are not normally for public disclosure)
- Summary of observations from the site evaluation
- Field safety screening results, including any appropriate caveats on the results
- Rapid field water, air, and/or soil testing results, including any appropriate caveats on the results
- Inventory of samples collected, and the sites from which they were collected
- Any other pertinent information developed during the site characterization

Sample Packaging and Transport

In order to perform analysis of samples beyond rapid field testing, it will be necessary to properly package the samples for transport to the appropriate labs as quickly as possible. Prompt and proper packaging and transport of samples will:

- Protect the integrity of samples from changes in composition or concentration caused by bacterial growth or degradation that might occur at increased temperatures
- Reduce the chance of leaking or breaking of sample containers that would result in loss of sample volume, loss of sample integrity, and potential exposure of personnel to hazardous substances
- Help ensure compliance with shipping regulations

REFERENCES AND RECOMMENDED READING

DHS (2010). Defense Industrial Base Infrastructure Specific Plan. Washington, DC: Department of Homeland Security.

Henry, K. (2002). New Face of Security. *Gov. Security*, pp. 30–37.

USDOE (2010). *Energy Sector-Specific Plan: An Annex to the National Infrastructure Protection Plan.* Washington, DC: USDOE.

USEPA (2001). *Protecting the Nation's Water Supplies from Terrorist Attack: Frequently Asked Questions.* Washington, DC: United States Environmental Protection Agency.

USEPA (2004). *Response Protocol Toolbox: Planning for and Responding to Drinking Water Contamination Threats and Incidents.* Washington, DC: United States Environmental Protection Agency.

USEPA (2011). *Response Protocol Toolbox: Planning for and Responding to Wastewater Contamination Threats and Incidents.* Washington, DC: United States Environmental Protection Agency.

7

Cybersecurity and the DIB Sector

> The use of networks and technical attacks against the resources connected to them is an evolution in the history of warfare.
>
> —Stiennon, *Surviving Cyberwar*

Just a cursory overview of reports of cyber exploits illustrate the potentially debilitating effects such attacks could and can have on the nation's security, economy, and on public health and safety.

- In May 2015, media sources reported that data belonging to 1.1 million health insurance customers in the Washington, DC, area were stolen in a cyberattack on a private insurance company. Attackers accessed a database containing names, birth dates, e-mail addresses, and subscriber ID numbers of customers.
- In December 2014, the industrial Control Systems Cyber Emergency Response Team (ICS-CERT; works to reduce risks within and across all critical infrastructure sectors by partnering with law enforce agencies) issued an updated alert on a sophisticated malware campaign compromising numerous industrial control system environments. Their analysis indicated that this campaign had been ongoing since at least 2011.
- In the January 2014 to April 2014 release of its Monitor report, ICS-CERT reported that a public utility had been compromised when a sophisticated

threat actor gained unauthorized access to its control system network through a vulnerable remote access capability configured on the system. The incident highlighted the need to evaluate security controls employed at the perimeter and ensure that potential intrusion vectors are configured with appropriate security controls, monitoring and detection capabilities.

In 2000, the FBI identified and listed threats to critical infrastructure. These threats are listed and described in table 7.1. In 2015, the GAO described the sources of cyber-based threats. These threats are listed and described in detail in table 7.2.

Table 7.1. Threats to Critical Infrastructure Observed by the FBI

Threat	*Description*
Criminal groups	There is an increased use of cyber intrusions by criminal groups who attack systems for purposes of monetary gain.
Foreign intelligence services	Foreign intelligence services use cyber tools as part of their information-gathering and espionage activities.
Hackers	Hackers sometimes crack into networks for the thrill of the challenge or for bragging rights in the hacker community. While remote cracking once required a fair amount of skill or computer knowledge, hackers can now download attack scripts and protocols from the Internet and launch them against victim sites. Thus, while attack tools have become more sophisticated, they have also become easier to use.
Hacktivists	Hacktivism refers to politically motivated attacks on publicly accessible Web pages or e-mail servers. These groups and individuals overload e-mail servers and hack into websites to send a political message.
Information warfare	Several nations are aggressively working to develop information warfare doctrine, programs, and capabilities. Such capabilities enable a single entity to have a significant and serious impact by disrupting the supply, communications, and economic infrastructures that support military power—impacts that, according to the Director of Central Intelligence, can affect the daily lives of Americans across the country.
Inside threat	The disgruntled organization insider is a principal source of computer crimes. Insiders may not need a great deal of knowledge about computer intrusions because their knowledge of a victim system often allows them to gain unrestricted access to cause damage to the system or to steal system data. The insider threat also includes outsourcing vendors.
Virus writers	Virus writers are posing an increasingly serious threat. Several destructive computer viruses and "worms" have harmed files and hard drives, including the Melissa Macro Virus, the Explore. Zip worm, the CIH (Chernobyl) Virus, Nimda, and Code Red.

FBI (2000, 2014).

Table 7.2. Common Cyber Threat Sources

Source	*Description*
Non-adversarial-malicious	
Failure in IT equipment	Failures in displays, sensors, controllers, and information technology hardware responsible for data storage, processing, and communications
Failure in environmental controls	Failures in temperature/humidity controllers or power supplies
Software coding errors	Failures in operating systems, networking, and general-purpose and mission-specific applications
Natural or man-made disaster	Events beyond an entity's control such as fires, floods/tsunamis, tornadoes, hurricanes, and earthquakes
Unusual or natural event	Natural events beyond the entity's control that are not considered to be disasters (e.g., sunspots)
Infrastructure failure or outage	Failure or outage of telecommunications or electrical power
Unintentional user errors	Failures resulting from erroneous, accidental actions taken by individuals (both system users and administrators) in the course of executing their everyday responsibilities
Adversarial	
Hackers or hacktivists	Hackers break networks for the challenge, revenge, stalking, or monetary gain, among other reasons. Hactivists are ideologically motivated actors who use cyber exploits to further political goals.
Malicious insiders	Insiders (e.g., disgruntled organization employees, including contractors) may not need a great deal of knowledge about computer intrusions because their position with the organization often allows them to gain unrestricted access and cause damage to the target system or to steal system data. These individuals engage in purely malicious activities and should not be confused with non-malicious insider accidents.
Nations	Nations, including nation-state, state-sponsored, and state-sanctioned programs use cyber tools as part of their information-gathering and espionage activities. In addition, several nations are aggressively working to develop information warfare doctrine, programs, and capabilities.
Criminal groups and organize crime	Criminal groups seek to attack systems for monetary gain. Specifically, organized criminal groups use cyber exploits to commit identity theft, online fraud, and computer extortion.

(continued)

Table 7.2. ***Continued***

Source	*Description*
Terrorist	Terrorists seek to destroy, incapacitate, or exploit critical infrastructures in order to threaten national security, cause mass casualties, weaken the economy, and damage public morale and confidence.
Unknown malicious outsiders	Unknown malicious outsiders are threat sources or agents that, due to a lack of information, agencies are unable to classify as being one of the five types of threat sources or agents listed above.

GAO Analysis of Unclassified Government and Nongovernmental Data. GAO 16-79.

Threats to systems supporting critical infrastructure are evolving and growing. As shown in table 7.2, cyber threats can be unintentional or intentional. Unintentional or non-adversarial threats include equipment failures, software coding errors, and the actions of poorly trained employees. They also include natural disasters and failures of critical infrastructure on which the organization depends but are outside of its control. Intentional threats include both targeted and untargeted attacks form a variety of sources, including criminal groups, hackers, disgruntled employees, foreign nations engaged in espionage and information warfare, and terrorists. These threat adversaries vary in terms of the capabilities of the actors, their willingness to act, and their motives, which can include seeking monetary gain or seeking an economic, political, or military advantage (GAO 2015).

Today's fast-evolving "information age" technology has intensified the importance of critical infrastructure protection, in which cybersecurity has become as critical as physical security. The Department of Defense (DoD) determined that cyber threats to contractors' unclassified information systems represented an unacceptable risk of compromise to DoD information and posed a significant risk to U.S. national security and economic security interests.

In the past few years, it has been somewhat routine for us to see a television news program or read an online media exposé where the focus is cybersecurity—or, more often, the *lack* of it. Recently, there has been discussion (and then indictments) about Russian hackers trying to influence the U.S. 2016 elections. Many of the cyber intrusion incidents we read or hear about have added new terms or new uses for old terms to our vocabulary. For example, old terms such as botnets (short for robot networks, also balled bots, zombies, botnet fleets, and many others)—groups of computers that have been compromised with malware such as Trojan horses, worms, backdoors, remote control software, and viruses—have taken on new connotations with regard to cybersecurity. Relatively new terms such as scanners, Windows NT hacking tools, ICQ hacking tools, mail bombs, sniffer, logic bomb, nukers, dots, backdoor Trojan, key loggers, hackers' Swiss knife, password crackers, blended threats, Warhol Worms, Flash Threats, Targeted Attacks, and BIOS crackers are now common. And new terms have evolved along with various control mechanisms. For example, because many control systems are vulnerable to attacks of varying degrees, these attack attempts range from telephone line sweeps (war dialing), to wireless network sniffing (war driving), to physical network port scanning, and to physical monitoring and intrusion. When wireless network sniffing is performed at (or near) the target point by a pedestrian, that is called warwalking, and it means that instead of a person being in vehicle, the potential intruder may be sniffing the network for weaknesses or vulnerabilities on foot, posing as a pedestrian, using a handheld PDA device or laptop computer (Warwalking 2003). Further, adversaries can leverage common computer software programs, such as Adobe Acrobat and Microsoft Office, to deliver a threat by embedding exploits within software files that can be activated when a user opens a file within its corresponding program. Table 7.3 provides descriptions of common exploits or techniques, tactics, and practices used by cyber adversaries.

Table 7.3. Common Methods of Cyber Exploits

Exploit	*Description*
Watering hole	A method by which threat actors exploit the vulnerabilities of carefully selected websites frequented by users of the targeted system. Malware is then injected to the targeted system via the compromised websites.
Phishing and spear phishing	A digital form of social engineering that uses authentic-looking e-mails, websites, or instant messages to get users to download malware, open malicious attachments, or open links that direct then to a website that requires information or executes malicious code.
Credentials based	An exploit that takes advantage of a system's insufficient user authentication and/or any elements of cybersecurity supporting it, to include not limiting the number of failed login attempts, the use of hard-coded credentials, and the use of a broken or risky cryptographic algorithm.
Trusted third parties	An exploit that takes advantage of the security vulnerabilities of trusted third parties to gain access to an otherwise secure system.
Classic buffer overflow	An exploit that involves the intentional transmission of more data than a program's input buffer can hold, leading to the deletion of critical data and subsequent execution of malicious code.
Cryptographic weakness	An exploit that takes advantage of a network employing insufficient encryption when either storing or transmitting data, enabling adversaries to read and/or modify the data stream.
Structured Query Language (SQL) injection	Involves the alteration of a database search in a web-based application, which can be used to obtain unauthorized across to sensitive information in a database, resulting in data loss at corruption, denial or service, or complete host takeover.
Operating system command injection	Takes advantage of a system's inability to properly neutralize special elements used in operating system commands, allowing the adversaries to execute unexpected commands on the system by either modifying already evoked commands or evoking their own.

Exploit	*Description*
Cross-site scripting	Uses third-party web resources to run lines of programming code (referred to as scripts) within the victim's web browser or scriptable application. This occurs when a user, using a browser, visits a malicious website or clicks a malicious link. The most dangerous consequences can occur when this method is used to exploit additional vulnerabilities that may permit an adversary to steal cookers (data exchanged between a web server and a browser), log key stokes, capture screen shots, discover and collect network information, or remotely access and control the victim's machine.
Cross-site request forgery	An exploit takes advantage of an application that cannot, or does not, sufficiently verify whether a well-formed, valid, consistent request was intentionally provided by the user who submitted the request, tricking the victim into executing a falsified requires that results in the system or data being compromised.
Path traversal	An exploit that seeks to gain access to files outside of a restricted directory by modifying the directory pathname in an application that does not properly neutralize special elements (e.g., ' . . . ', '/', ' . . . /', etc.)
Integer overflow	An exploit where malicious code is inserted that leads to unexpected integer overflow, or wraparound, which can be used by adversaries to control looping or make security decisions in order to cause program crashes, memory corruption, or the execution of arbitrary code via buffer overflow.
Uncontrolled format string	Adversaries manipulate externally controlled format strings in print-style functions to gain access to information and/or execute unauthorized code or commands.
Open redirect	An exploit where the victim is tricked into selecting a URL (website location) that has been modified t direct them to an external, malicious site which may contain malware that can compromises the victim's machine.

(*continued*)

Table 7.3. *Continued*

Exploit	*Description*
Heap-based buffer overflow	Similar to classic buffer overflow, but the buffer that is overwritten is allocated in the heap portion of memory, generally meaning that the buffer was allocated using a memory allocation routine, such as "malloc ( )".
Unrestricted upload of files	An exploit that takes advantage of insufficient upload restrictions, enabling adversaries to upload malware (e.g., .php) in place of the intended file type (e.g., .jpg).
Inclusion of functionality from untrusted sphere	Uses trusted, third-party executable functionality (e.g., web widget or library) as a means of executing malicious code in software whose protection mechanism are unable to determent whether functionality a from a trusted sources, modified in transit, or being spoofed.
Certificate and certificate authority compromise	Exploits facilitated via the issuance of fraudulent digital certificates (e.g., transport layer security and Secure Socket Layer). Adversaries use these certificates to establish secure connections with the target organization or individual by mimicking a trusted third party.
Hybrid of others	An exploit which combines elements of two or more of the aforementioned techniques.

GAO (2015).

WHAT IS SCADA?

SCADA is a digital term that most people have not heard of. This is not the case, however, for those who work with the nation's critical infrastructure, including DIB sector infrastructure. SCADA, or **S**upervisory **C**ontrol **a**nd **D**ata **A**cquisition System (also sometimes referred to as Digital Control Systems or Process Control Systems), plays an important role in computer-based control systems. From coordinating music and lights in the proper sequence with spray from water fountains to controlling systems used in the drilling and refining of oil and natural gas, control systems perform many functions. Many energy distribution networks use computer-based systems to remotely control sensitive feeds, processes, and system equipment previously controlled manually. These systems allow an energy utility (or operation) to monitor fuel tank levels, to ensure that contents are stored at correct levels or to monitor tank levels, and, as mentioned, collect data from sensors and control equipment located at remote sites. Common DIB sector system sensors measure many parameters, depending on the nature of the industrial manufacturing entity. Common DIB industry system equipment includes valves, pumps, and switching devices for distribution of electricity. The critical infrastructure of many countries is increasingly dependent on SCADA systems.

If we were to ask the specialist to define SCADA, the technical response could be outlined as follows:

SCADA

- is a multi-tier system or interfaces with multi-tier systems
- is used for physical measurement and control endpoints via an Remote Terminal Unit (RTU) and Programmable Logic Controller (PLC) to measure voltage, adjust a value, or flip a switch
- is an intermediate processor normally based on commercial third-party OSs—VMS, Unix, Windows, Linux
- is human interfaces, for example, with graphical user interface (Windows GUIs)
- has a communication infrastructure that consists of a variety of transport mediums such as analog, serial, Internet, radio, Wi-Fi

But how about the nonspecialist response?—for the rest of us who are nonspecialists. Well, for the nonspecialist and the rest of us we could simply say that SCADA is a computer-based control system that remotely controls and monitors processes previously controlled and monitored manually. The philosophy behind SCADA control systems can be summed up by the phrase "if you can measure it, you can control it." SCADA allows an operator using a central computer to supervise (control and monitor) multiple networked computers at remote locations. Each remote computer can control mechanical processes (mixers, pumps, valves, etc.) and collect data from sensors at its remote location. Thus the phrase: **S**upervisory **C**ontrol **a**nd **D**ata **A**cquisition, or **SCADA**.

The central computer is called the Master Terminal Unit, or MTU. The MTU has two main functions: periodically obtain data from RTUs/PLCs and control remote devices through the operator station. The operator interfaces with the MTU using software called Human Machine Interface, or HMI. The remote computer is called Programmable Logic Controller (PLC) or Remote Terminal Unit (RTU). The RTU activates a relay (or switch) that turns mechanical equipment "on" and "off." The RTU also collects data from sensors. Sensors perform measurement, and actuators perform control.

In the initial stages utilities ran wires, also known as hardwire or land lines, from the central computer (MTU) to the remote computers (RTUs). Because remote locations can be located hundreds of miles from the central location, utilities began to use public phone lines and modems, leased telephone company lines, and radio and microwave communication. More recently, they have also begun to use satellite links, Internet, and newly developed wireless technologies.

DID YOU KNOW?

Modern RTUs typically use a ladder-logic approach to programming due to its similarity to standard electrical circuits. An RTU that employees this ladder logic programming is called a Programmable Logic Controller (PLC).

Because SCADA systems' sensors provided valuable information, many critical infrastructure entities, utilities, and other industries established "connections" between their SCADA systems and their business system. This allowed utility/industrial management and other staff access to valuable statistics, such as chemical usage. When utilities/industries later connected their systems to the Internet, they were able to provide stakeholders/stockholders with usage statistics on the DIB segment and utility/industrial web pages. Figure 7.1 provides a basic illustration of a representative SCADA network. Note that firewall protection would normally be placed between Internet and business system and between business system and the MTU.

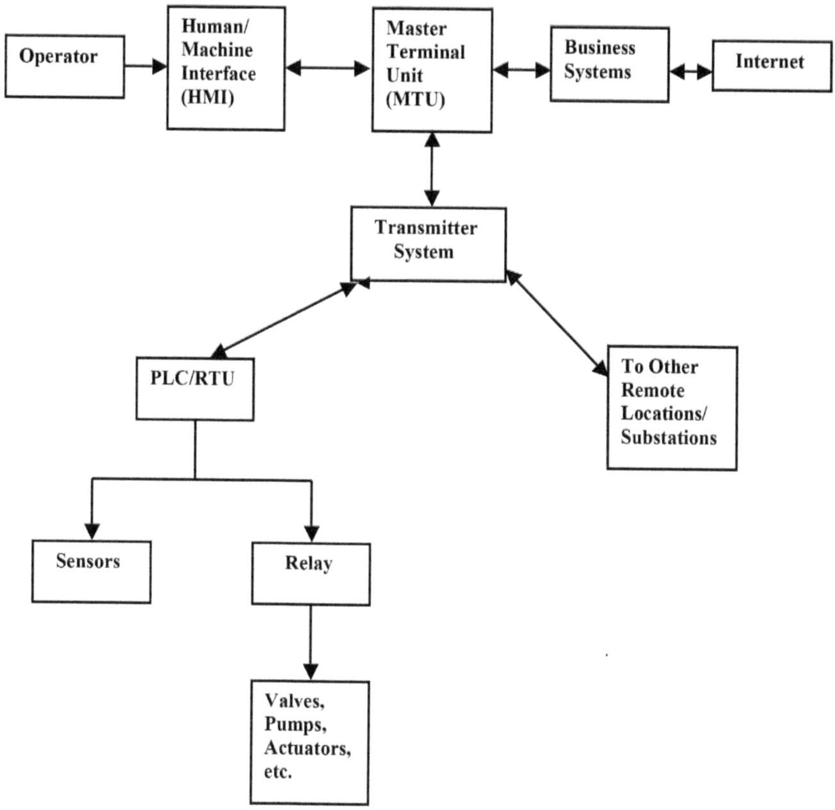

FIGURE 7.1
Representative SCADA Network

SCADA Applications

As stated previously, SCADA systems can be designed to measure a variety of equipment operating conditions; to measure parameters or volumes and flow rates of electricity, natural gas, and oil or oil and petrochemical mixture quality; and to respond to changes in those parameters either by alerting operators or by modifying system operation through a feedback loop system—without having personnel physically visit each valve, process, or piece of other equipment on a daily basis to check it and/or ensure that it is functioning properly. Automation and integration of large-scale diverse assets required SCADA systems to provide the utmost in flexibility, scalability, openness, and reliability. SCADA systems are used to automate certain energy production functions; these can be performed without initiation by an operator. In addition to process equipment, SCADA systems can also integrate specific security alarms and equipment, such as cameras, motion sensors, lights, data from card reading systems, and so on, thereby providing a clear picture of what is happening at areas throughout a facility. Finally, SCADA systems also provide constant, real-time data on processes, equipment, location access, and so on, for the necessary response to be made quickly. This can be extremely useful during emergency conditions, such as when energy distribution lines or piping breaks or when potentially disruptive chemical reaction spikes appear in chemical processing operations. Currently, it can be said that SCADA has evolved from a simple indicating light and pushbutton control system into a comprehensive operation and handling system for very complex process control and safety shutdown systems. In a nutshell, SCADA results in an oversight system that requires fewer operators.

Today, many common DIB sector applications for SCADA systems include, but are not limited to those shown here:

- Boiler controls
- Bearing temperature monitor (electric generators and motors)
- Gas processing
- Plant monitoring
- Plant energy management
- Power distribution monitoring
- Electric power monitoring
- Fuel oil handling system
- Hydroelectric load management
- Petroleum pilot plants
- Plant monitoring
- Process controls
- Process stimulators
- Tank controls
- Utility monitoring
- Safety parameter display systems and shutdown systems
- Tank level control and monitoring
- Turbine controls

- Turbine monitoring
- Virtual annunciator panels
- Alarm systems
- Security equipment
- Event logging

Because these systems can monitor multiple processes, equipment, and infrastructure and then provide quick notification of, or response to, problems or upsets, SCADA systems typically provide the first line of detection for atypical or abnormal conditions. For example, a SCADA system connected to sensors that measure specific machining quality parameters are measured outside of a specific range. A real-time customized operator interface screen could display and control critical systems monitoring parameters.

The system could transmit warning signals back to the operators, such as by initiating a call to a personal pager. This might allow the operators to initiate actions to prevent power outages or contamination and disruption of the energy supply. Further automation of the system could ensure that the system initiated measures to rectify the problem. Preprogrammed control functions (e.g., shutting a valve, controlling flow, throwing a switch, or adding chemicals) can be triggered and operated based on SCADA utility.

SCADA Vulnerabilities

Even though terrorists, domestic and/or foreign, tend to aim their main focus around the critical devices that control actual DIB production and delivery activities, according to USEPA (2005), SCADA networks were originally developed with little attention paid to security, making the security of these systems often weak. Studies have found that, while technological advancements introduced vulnerabilities, many DIB sector plans/sites and utilities have spent little time securing their SCADA networks. As a result, many SCADA networks may be susceptible to attacks and misuse. SCADA systems languished in obscurity and this was the essence of their security; that is, until technological developments transformed SCADA from a backroom operation to a front-and-center visible control system.

Remote monitoring and supervisory control of processes began to develop in the early 1960s, and adopted many technological advancements. The advent of minicomputers made it possible to automate a vast number of once

U.S. ELECTRIC GRID IS HACKED INTO

The Associated Press (AP) reported on April 9, 2009, that spies hacked into the U.S. energy grid and left behind computer programs (Trojan horses) that would enable them to disrupt service, exposing potentially catastrophic vulnerabilities in key pieces of national infrastructure.

manually operated switches. Advancements in radio technology reduced the communication costs associated with installing and maintaining buried cable in remote areas. SCADA systems continued to adopt new communication methods including satellite and cellular. As the price of computers and communications dropped, it became economically feasible to distribute operations and to expand SCADA networks to include even smaller facilities.

Advances in information technology and the necessity of improved efficiency have resulted in increasingly automated and interlinked infrastructures, and created new vulnerabilities due to equipment failure, human error, weather and other natural causes, and physical and cyberattacks. Some areas and examples of possible SCADA vulnerabilities include (Wiles et al. 2007):

- **Human**—People can be tricked or corrupted and may commit errors.
- **Communications**—Messages can be fabricated, intercepted, changed, deleted, or blocked.
- **Hardware**—Security features are not easily adapted to small self-contained units with limited power supplies.
- **Physical**—Intruders can break into a facility to steal or damage SCADA equipment.
- **Natural**—Tornadoes, floods, earthquakes, and other natural disasters can damage equipment and connections.
- **Software**—Programs can be poorly written.

Specific SCADA weaknesses and potential attack vectors include:

- Does not require any authorization.
- Does not use encryption.
- Does not properly recognize and properly handle errors and exceptions.
- No authentication is required.
- Data could be intercepted.
- Manipulation of data.
- Service denial.
- IP address spoofing—Internet protocol packets with a false source IP address.
- Session hijacking.
- Unsolicited responses
- Packet fuzzing—involves inputting false information, etc.
- Unauthorized control

SCADA system computers and their connections are susceptible to different types of information system attacks and misuse such as those mentioned above. The Computer Security Institute and Federal Bureau of Investigation conduct an annual Computer Crime and Security Survey (FBI 2007). The survey reported on 10 types of attacks or misuse and reported that virus and denial of service had the greatest negative economic impact. The same study also found that 15 percent of the respondents reported abuse of wireless networks, which can be a SCADA component. On average, respondents from all sectors did not believe that their organization invested enough in security awareness. Utilities as a group reported a lower average computer security expenditure/investment per employee than many other sectors such as transportation, telecommunications, and financial.

Sandia National Laboratories' *Common Vulnerabilities in Critical Infrastructure Control Systems* described some of the common problems it has identified in the following five categories (Stamp et al., 2003):

1. **System Data**—Important data attributes for security include availability, authenticity, integrity, and confidentiality. Data should be categorized according to its sensitivity, and ownership and responsibility must be assigned.

However, SCADA data is often not classified at all, making it difficult to identify where security precautions are appropriate (for example, which communication links to secure, databases requiring protection, etc.).

2. **Security Administration**—Vulnerabilities emerge because many systems lack a properly structured security policy (security administration is notoriously lax in the case of control systems), equipment and system implementation guides, configuration management, training, and enforcement and compliance auditing.
3. **Architecture**—Many common practices negatively affect SCADA security. For example, while it is convenient to use SCADA capabilities for other purposes such as fire and security systems, these practices create single points of failure. Also, the connection of SCADA networks to other automation systems and business networks introduces multiple entry points for potential adversaries.
4. **Network** (including communication links)—Legacy systems' hardware and software have very limited security capabilities, and the vulnerabilities of contemporary systems (based on modern information technology) are publicized. Wireless and shared links are susceptible to eavesdropping and data manipulation.
5. **Platforms**—Many platform vulnerabilities exist, including default configurations retained, poor password practices, shared accounts, inadequate protection for hardware, and nonexistent security monitoring controls. In most cases, important security patches are not installed, often due to concern about negatively impacting system operation; in some cases technicians are contractually forbidden from updating systems by their vendor agreements.

The following incident helps to illustrate some of the risks associated with SCADA vulnerabilities. During the course of conducting a vulnerability assessment, a contractor stated that personnel from his company penetrated the information system of a utility within minutes. Contractor personnel drove to a remote substation and noticed a wireless network antenna. Without leaving their vehicle, they plugged in their wireless radios and connected to the network within five minutes. Within 20 minutes they had mapped the network, including SCADA equipment, and accessed the business network and data. This illustrates what a cybersecurity advisor from Sandia National Laboratories specializing in SCADA stated, that utilities are moving to wireless communication without understanding the added risks.

The Increasing Risk

According to GAO (2015), historically, security concerns about control systems (SCADA included) were related primarily to protecting against physical attack and misuse of refining and processing sites or distribution and holding facilities. However, more recently there has been a growing recognition that control systems are now vulnerable to cyberattacks from numerous sources, including hostile governments, terrorist groups, disgruntled employees, and other malicious intruders.

In addition to the vulnerabilities mentioned earlier, several factors have contributed to the escalation of risk to control systems, including (1) the adoption of standardized technologies with known vulnerabilities, (2) the connectivity of control systems to other networks, (3) constraints on the implementation of existing security technologies and practices, (4) insecure remote connections, and (5) the widespread availability of technical information about control systems.

Adoption of Technologies with Known Vulnerabilities

When a technology is not well known, not widely used, or not understood or publicized, it is difficult to penetrate it and thus disable it. Historically, proprietary hardware, software, and network protocols made it difficult to understand how control systems operated—and therefore how to hack into them. Today, however, to reduce costs and improve performance, organizations have been transitioning from proprietary systems to less expensive, standardized technologies such as Microsoft's Windows and Unix-like operating systems and the common networking protocols used by the Internet. These widely used standardized technologies have commonly known vulnerabilities, and sophisticated and effective exploitation tools are widely available and relatively easy to use. As a consequence, both the number of people with the knowledge to wage attacks and the number of systems subject to attack have increased. Also, common communication protocols and the emerging use of Extensible Markup Language (commonly referred to as XML) can make it easier for a hacker to interpret the content of communications among the components of a control system.

Control systems are often connected to other networks—enterprises often integrate their control system with their enterprise networks. This increased connectivity has significant advantages, including providing decision-makers with access to real-time information and allowing engineers to monitor and control the process control system from different points on the enterprise network. In addition, the enterprise networks are often connected to the networks of strategic partners and to the Internet. Further, control systems are increasingly using wide area networks and the Internet to transmit data to their remote or local stations and individual devices. This convergence of control networks with public and enterprise networks potentially exposes the control systems to additional security vulnerabilities. Unless appropriate security controls are deployed in the enterprise network and the control system network, breaches in enterprise security can affect the operation of controls system.

According to industry experts, the use of existing security technologies, as well as strong user authentication and patch management practices, are generally not implemented in control systems because control systems operate in real time, typically are not designed with cybersecurity in mind, and usually have limited processing capabilities.

Existing security technologies such as authorization, authentication, encryption, intrusion detection, and filtering of network traffic and communications require more bandwidth, processing power, and memory than control system components typically have. Because controller stations are generally designed to do specific tasks, they use low-cost, resource-constrained microprocessors. In fact, some devices in the electrical industry still use the Intel 8088 processor, introduced in 1978. Consequently, it is difficult to install existing security technologies without seriously degrading the performance of the control system.

Further, complex passwords and other strong password practices are not always used to prevent unauthorized access to control systems, in part because this could hinder a rapid response to safety procedures during an emergency. As a result, according to experts, weak passwords that are easy to guess, shared, and infrequently change are reportedly common in control systems, including the use of default passwords or *even no password at all.*

In addition, although modern control systems are based on standard operating systems, they are typically customized to support control system applications. Consequently, vendor-provided software patches are generally either incompatible or cannot be implemented without compromising service shutting down "always-on" systems or affecting interdependent operations.

Potential vulnerabilities in control systems are exacerbated by insecure connections. Organizations often leave access links—such as dial-up modems to equipment and control information—open for remote diagnostics, maintenance, and examination of system status. Such links may not be protected with authentication of encryption, which increases the risk that hackers could us these insecure connections to break into remotely controlled systems. Also, control systems often use wireless communications systems, which are especially vulnerable to attack, or leased lines that pass through commercial telecommunications facilities. Without encryption to protect data as it flows through these insecure connections or authentication mechanisms to limit access, there is limited protection for the integrity of the information being transmitted.

Public information about infrastructures and control systems is available to potential hackers and intruders. The availability of this infrastructure and vulnerability data was demonstrated by a university graduate student whose dissertation reportedly mapped every business and industrial sector in the American economy to the fiber optic network that connects them—using material that was available publicly on the Internet, none of which was classified.

Cyber Threats to Control Systems

As components of control systems increasingly make critical decisions that were once made by humans, the potential effect of a cyber threat becomes more devastating.

In July 2002, National Infrastructure Protection Center (NIPC) reported that the potential for compound cyber and physical attacks, referred to as "swarming attacks," was an emerging threat to the U.S. critical infrastructure. As NIPC reports, the effects of a swarming attack include slowing or complicating the response to a physical attack. For instance, a cyberattack that disables the water supply or the electrical system in conjunction with a physical attack could deny emergency services the necessary resources to manage the consequences—such as controlling fires, coordinating actions, and generating light.

Entities or individuals with malicious intent might take one or more of the following actions to successfully attack SCADA/control systems:

- Disrupt the operation of control systems by delaying or blocking the flow of information through control networks, thereby denying availability of the networks to control system operations
- Make unauthorized changes to programmed instructions in PLCs, RTUs, or DCS controllers, change alarm thresholds, or issue unauthorized commends to control equipment, which could potentially result in damage to equipment (if tolerances are exceeded), premature shutdown of processes (such as prematurely shutting down transmission lines), or even disabling of control equipment
- Send false information to control system operators either to disguise unauthorized changes or to initiate inappropriate actions by system operators
- Modify the control system software, producing unpredictable results
- Interfere with the operation of safety systems

In addition, in control systems that cover a wide geographic area, the remote sites are often unstaffed and may not be physically monitored. If such remote systems are physically breached, the attackers could establish a cyber connection to the control network.

Securing Control Systems

Several challenges must be addressed to effectively secure control systems against cyber threats. These challenges include: (1) the limitations of current security technologies in securing control systems; (2) the perception that securing control systems may not economically justifiable; and (3) the conflicting priorities within organizations regarding the security of control systems.

A significant challenge in effectively securing control systems is the lack of specialized security technologies for these systems. The computing resources in control systems that are needed to perform security functions tend to be quite limited, making it very difficult to use security technologies within control system networks without severely hindering performance.

Securing control systems may not be perceived as economically justifiable. Experts and industry representatives have indicated that organizations may be reluctant to spend more money to secure control systems. Hardening the security of control systems would require industries to expend more resources, including acquiring more personnel, providing training for personnel, and potentially prematurely replacing current systems that typically have a lifespan of about twenty years.

Finally, several experts and industry representatives indicated that the responsibility for securing control systems typically includes two separate groups: IT security personnel and control system engineers and operators. IT security personnel tend to focus on securing enterprise systems, while control system engineers and operators tend to be more concerned with the reliable performance of their control systems. Further, they indicate that, as a result, those two groups do not always fully understand each other's requirements and collaborate to implement secure control systems.

STEPS TO IMPROVE SCADA SECURITY

The President's Critical Infrastructure Protection Board and the Department of Energy (DOE) have developed the steps outlined below to help organizations improve the security of their SCADA networks. DOE (2001) points out that these steps are not meant to be prescriptive or all-inclusive. However, they do address essential actions to be taken to improve the protection of SCADA networks. The steps are divided into two categories: specific actions to improve implementation, and actions to establish essential underlying management processes and policies.

21 Steps to Increase SCADA Security

The following steps focus on specific actions to be taken to increase the security of SCADA networks (DOE 2001):

1. **Identify all connections to SCADA networks.**

   Conduct a thorough risk analysis to assess the risk and necessity of each connection to the SCADA network. Develop a comprehensive understanding of all connections to the SCADA network, and how well those connections are protected. Identify and evaluate the following types of connections:

   - Internal local area and wide area networks, including business networks
   - The Internet
   - Wireless network devices, including satellite uplinks
   - Modem or dial-up connections
   - Connections to business partners, vendors, or regulatory agencies

2. **Disconnect unnecessary connections to the SCADA network.**
To ensure the highest degree of security of SCADA systems, isolate the SCADA network from other network connections to as great a degree as possible. Any connection to another network introduces security risks, particularly if the connection creates a pathway from or to the Internet. Although direct connections with other networks may allow important information to be passed efficiently and conveniently, insecure connections are simply not worth the risk; isolation of the SCADA network must be a primary goal to provide needed protection. Strategies such as utilization of "demilitarized zones" (DMZs) and data warehousing can facilitate the secure transfer of data from the SCADA network to business networks. However, they must be designed and implemented properly to avoid introduction of additional risk through improper configuration.
3. **Evaluate and strengthen the security of any remaining connections to the SCADA networks.**
Conduct penetration testing or vulnerability analysis of any remaining connections to the SCADA network to evaluate the protection posture associated with these pathways. Use this information in conjunction with risk management processes to develop a robust protection strategy for any pathways to the SCADA network. Since the SCADA network is only as secure as its weakest connecting point, it is essential to implement firewalls, intrusion-detection systems (IDSs), and other appropriate security measures at each point of entry. Configure firewall rules to prohibit access from and to the SCADA network, and be as specific as possible when permitting approved connections. For example, an independent system operator (ISO) should not be granted "blanket" network access simply because there is a need for a connection to certain components of the SCADA system. Strategically place IDSs at each entry point to alert security personnel of potential breaches of network security. Organization management must understand and accept responsibility for risks associated with any connection to the SCADA network.
4. **Harden SCADA networks by removing or disabling unnecessary services.**
SCADA control servers built on commercial or open-source operating systems can be exposed to attack default network services. To the greatest degree possible, remove or disable unused services and network demons

to reduce the risk of direct attack. This is particularly important when SCADA networks are interconnected with other networks. Do not permit a service or feature on a SCAA network unless a thorough risk assessment of the consequences of allowing the service/feature shows that the benefits of the service/feature far outweigh the potential for vulnerability exploitation. Examples of services to remove from SCADA networks include automated meter reading/remote billing systems, email services, and Internet access. An example of a feature to disable is remote maintenance. Numerous secure configurations for both commercial and open source operating **systems** are in the public domain, such as the National Security Agency's series of security guides. Additionally, work closely with SCADA vendors to identify secure configurations and coordinate any and all changes to operational systems to ensure that removing or disabling services does not cause downtime, interruption of service, or loss of support.

5. **Do not rely on proprietary protocols to protect your system.** Some SCADA systems are unique, proprietary protocols for communications between field devices and servers. Often the security of SCADA systems is based solely on the secrecy of these protocols. Unfortunately, obscure protocols provide very little "real" security. Do not rely on proprietary protocols or factor default configuration setting to protect your system. Additionally, demand that vendors disclose any backdoors or vendor interfaces to your SCADA systems, and expect them to provide systems that are capable of being secured.
6. **Implement the security features provided by device and system vendors.** Older SCADA systems (most systems in use) have no security features whatsoever, so SCADA system owners must insist that their system vendor implement security features in the form of product patches or upgrades. Some newer SCADA devices are shipped with basic security features, but these are usually disabled to ensure ease of installation.

   Analyze each SCADA device to determine whether security features are present. Additionally, factory default security settings (such as in computer network firewalls) are often set to provide maximum usability, but minimal security. Set all security features to provide the maximum security only after a thorough risk assessment of the consequences of reducing the security level.

7. **Establish strong controls over any medium that is used as a backdoor into the SCADA network.**
Where backdoors or vendor connections do exist in SCADA systems, strong authentication must be implemented to ensure secure communications. Modems, wireless, and wired networks used for communications and maintenance represent a significant vulnerability to the SCADA network and remote sites. Successful "war dialing" or "war driving" attacks would allow an attacker to bypass all other controls and have direct access to the SCADA network or resources. To minimize the risk of such attacks, disable inbound access and replace it with some type of callback system.
8. **Implement internal and external intrusion-detection systems and establish 24-hour-a-day incident monitoring.**
To be able to effectively respond to cyberattacks, establish an intrusion-detection strategy that includes alerting network administrators of malicious network activity originating from internal or external sources. Intrusion-detection system monitoring is essential 24 hours a day; this capability can be easily set up through a pager. Additionally, incident-response procedures must be in place to allow an effective response to any attack. To complement network monitoring, enable logging on all systems and audit system logs daily to detect suspicious activity as soon as possible.
9. **Perform technical audits of SCADA devices and networks, and any other connected networks, to identify security concerns.**
Technical audits of SCADA devices and networks are critical to ongoing security effectiveness. Many commercial and open-sourced security tools are available that allow system administrators to conduct audits of their systems/networks to identify active services, patch level, and common vulnerabilities. The use of these tools will not solve systemic problems but will eliminate the "paths of least resistance" that an attacker could exploit. Analyze identified vulnerabilities to determine their significance, and take corrective actions as appropriate. Track corrective actions and analyze this information to identify trends. Additionally, retest systems after corrective actions have been taken to ensure that vulnerabilities were actually eliminated. Scan non-production environments actively to identify and address potential problems.
10. **Conduct physical security surveys and assess all remote sites connected to the SCADA network to evaluate their security.**

Any location that has a connection to the SCADA network is a target, especially unmanned or unguarded remote sites. Conduct a physical security survey and inventory access points at each facility that has a connection to the SCADA system. Identify and assess any source of information including remote telephone/computer network/fiber optic cables that could be tapped; radio and microwave links that are exploitable; computer terminals that could be accessed; and wireless local area network access points. Identify and eliminate single points of failure. The security of the site must be adequate to detect or prevent unauthorized access. Do not allow "live" network access points at remote, unguarded sites simply for convenience.

11. **Establish SCADA "Red Teams" to identify and evaluate possible attack scenarios.**
    Establish a "Red Team" to identify potential attack scenarios and evaluate potential system vulnerabilities. Use a variety of people who can provide insight into weaknesses of the overall network, SCADA system, physical systems, and security controls. People who work on the system every day have great insight into the vulnerabilities of your SCADA network and should be consulted when identifying potential attack scenarios and possible consequences. Also, ensure that the risk from a malicious insider is fully evaluated, given that this represents one of the greatest threats to an organization. Feed information resulting from the "Red Team" evaluation into risk management processes to assess the information and establish appropriate protection strategies.

The following steps focus on management actions to establish an effective cybersecurity program:

12. **Clearly define cybersecurity roles, responsibilities, and authorities for managers, system administrators, and users.**
    Organization personnel need to understand the specific expectations associated with protecting information technology resources through the definition of clear and logical roles and responsibilities. In addition, key personnel need to be given sufficient authority to carry out their assigned responsibilities. Too often, good cybersecurity is left up to the initiative of the individual, which usually leads to inconsistent implementations

and ineffective security. Establish a cybersecurity organizational structure that defines roles and responsibilities and clearly identifies how cybersecurity issues are escalated and who is notified in an emergency.

13. **Document network architecture and identify systems that serve critical functions or contain sensitive information that require additional levels of protection.**

    Develop and document robust information security architecture as part of a process to establish an effective protection strategy. It is essential that organizations design their network with security in mind and continue to have a strong understanding of their network architecture throughout its lifecycle. Of particular importance, an in-depth understanding of the functions that the systems perform and the sensitivity of the stored information is required. Without this understanding, risk cannot be properly assessed and protection strategies may not be sufficient. Documenting the information security architecture and its components is critical to understanding the overall protection strategy, and identifying single points of failure.

14. **Establish a rigorous, ongoing risk management process.**

    A thorough understanding of the risks to network computing resources from denial-of-service attacks and the vulnerability of sensitive information to compromise is essential to an effective cybersecurity program. Risk assessments form the technical basis of this understanding and are critical to formulating effective strategies to mitigate vulnerabilities and preserve the integrity of computing resources. Initially, perform a baseline risk analysis based on current threat assessment to use for developing a network protection strategy. Due to rapidly changing technology and the emergence of new threats on a daily basis, an ongoing risk assessment process is also needed so that routine changes can be made to the protection strategy to ensure it remains effective. Fundamental to risk management is identification of residual risk with a network protection strategy in place and acceptance of that risk by management.

15. **Establish a network protection strategy based on the principle of defense-in-depth.**

    A fundamental principle that must be part of any network protection strategy is defense-in-depth. Defense-in-depth must be considered early in the design phase of the development process, and must be an integral consideration in all technical decision-making associated with the network. Utilize technical and administrative controls to mitigate threats from identified risks to as great a degree as possible at all levels of the network. Single points of failure must be avoided, and cybersecurity defense must be layered to limit and contain the impact of any security incidents. Additionally, each layer must be protected against other systems at the same layer. For example, to protect against the inside threat, restrict users to access only those resources necessary to perform their job functions.

16. **Clearly identity cybersecurity requirements.**
Organizations and companies need structured security programs with mandated requirements to establish expectations and allow personnel to be held accountable. Formalized policies and procedures are typically used to establish and institutionalize a cybersecurity program. A formal program is essential of establishing a consistent, standards-based approach to cybersecurity through an organization and eliminates sole dependence on individual initiative. Policies and procedures also inform employees of their specific cybersecurity responsibilities and the consequences of failing to meet those responsibilities. They also provide guidance regarding actions to be taken during a cybersecurity incident and promote efficient and effective actions during a time of crisis. As part of identifying cybersecurity requirements, include user agreements and notification and warning banners. Establish requirements to minimize the threat from malicious insiders, including the need for conducting background checks and limiting network privileges to those absolutely necessary.

17. **Establish effective configuration management processes.**
A fundamental management process needed to maintain a secure network is configuration management. Configuration management needs to cover both hardware configurations and software configurations. Changes to hardware or software can easily introduce vulnerabilities that undermine network security. Processes are required to evaluate and control any change to ensure that the network remains secure. Configuration management begins with well-tested and documented security baselines for your various systems.

18. **Conduct routine self-assessments.**
Robust performance evaluation processes are needed to provide organizations with feedback on the effectiveness of cybersecurity policy and

technical implementation. A sign of a mature organization is one that is able to self-identify issues, conduct root cause analyses, and implement effective corrective actions that address individual and systemic problems. Self-assessment processes that are normally part of an effective cybersecurity program include routine scanning for vulnerabilities, automated auditing of the network, and self-assessments of organizational and individual performance.

19. **Establish system backups and disaster recovery plans.**
Establish a disaster recovery plan that allows for rapid recover from any emergency (including a cyberattack). System backups are an essential part of any plan and allow rapid reconstruction of the network. Routinely exercise disaster recovery plans to ensure that they work and that personnel are familiar with them. Make appropriate changes to disaster recovery plans based on lessons learned from exercises.

20. **Senior organizational leadership should establish expectations for cybersecurity performance and hold individuals accountable for their performance.**
Effective cybersecurity performance requires commitment and leadership from senior managers in the organization. It is essential that senior management establish an expectation for strong cybersecurity and communicate this to their subordinate managers throughout the organization. It is also essential that senior organizational leadership establish a structure for implementation of a cybersecurity program. This structure will promote consistent implementation and the ability to sustain a strong cybersecurity program. It is then important for individuals to be held accountable for their performance as it relates to cybersecurity. This includes managers, system administrators, technicians, and users/operators.

21. **Establish policies and conduct training to minimize the likelihood that organizational personnel will inadvertently disclose sensitive information regarding SCADA system design, operations, or security controls.**
Release data related to the SCADA network only on a strict, need-to-know basis, and only to persons explicitly authorized to receive such information. "Social engineering," the gathering of information about a computer or computer network via questions to naïve users, is often the first step in a malicious attack on computer networks. The more information re-

vealed about a computer or computer network, the more vulnerable the computer/network is. Never divulge data revealed to a SCADA network, including the names and contact information about the system operators/administrators, computer operating systems, and/or physical and logical locations of computers and network systems over telephones or to personnel unless they are explicitly authorized to receive such information. Any requests for information by unknown persons need to be sent to a central network security location for verification and fulfillment. People can be a weak link in an otherwise secure network. Conduct training and information awareness campaigns to ensure that personnel remain diligent in guarding sensitive network information, particularly their passwords.

REFERENCES AND RECOMMENDED READING

Associated Press (AP) (2009). *Goal: Disrupt.* From the 04/04/09 The Virginian-Pilot, Norfolk, Va.

Brown, A.S. (2008). *SCADA vs. The Hackers.* American Society of Mechanical Engineers. Accessed 05/10/08 at http://www.memagazine.org/backissues/dec02/features/scadavs/.

DOE (2001). *21 Steps to improve cyber security of SCADA networks.* Washington, DC: Department of Energy.

Ezell, B.C. (1998). *Risks of Cyber Attack to Supervisory Control and Data Acquisition.* VA: University of Virginia.

FEMA (2008). *FEMA452: Risk Assessment, A How-to Guide.* Accessed 05/01/08 at fema.gov/library/file?type=published/filetofile.

FEMA (2015). *Protecting Critical Infrastructure against Insider Threats.* Accessed 17 April 2015 at http://emilms.fema.gov/IS0915/IABsummary.htm.

FBI (2000). *Threat to Critical Infrastructure.* Washington, DC: Federal Bureau of Investigation.

FBI (2007). *Ninth Annual Computer Crime and Security Survey.* FBI: Computer Crime Institute and Federal Bureau of Investigation.

FBI (2014). *Protecting Critical Infrastructure and the Importance of Partnerships.* Accessed at https://www.fbi.gov/news/speeches/protecting-critical-infrastructure-and-the-importance-of-partnerships.

GAO (2003). *Critical Infrastructure Protection: Challenges in Securing Control System.* Washington, DC: United States General Accounting Office.

GAO (2015). *Critical Infrastructure Protection: Sector-Specific Agencies Need to Better Measure Cybersecurity Progress*. Washington, DC: United States Government Accountability Office.

National Infrastructure Advisory Council (2008). *First Report and Recommendations on the Insider Threat to Critical Infrastructure.* Washington, DC.

NIPC (2002). *National Infrastructure Protection Center Report.* Washington, DC: National Infrastructure Protection Center.

Stamp, J. et al. (2003). *Common Vulnerabilities in Critical Infrastructure Control Systems*, 2nd ed. Sandia National Laboratories.

Stiennon, R. (2010). *Surviving Cyberwar.* Lanham, MD: Rowman & Littlefield Publishing Group.

USDOE (2010). *Energy Sector-Specific Plan: An Annex to the National Infrastructure Protection Plan.* Washington, DC: USDOE.

USDOE (2002). *Vulnerability Assessment Methodology: Electric Power Infrastructure.* Washington, DC.

USEPA (2005). EPA Needs to Determine What Barriers Prevent Water Systems from Securing Known SCADA Vulnerabilities. In: Harris, J. Final Briefing Report. Washington, DC: USEPA.

Warwalking (2003). Accessed 05/09/08 at http://warwalking.tribe.net.

Wiles, J. et al. (2007). *Techno Security's™ Guide to Securing SCADA* . Burlington, MA: Elsevier, Inc.

Young, M.A. (2004). *SCADA Systems Security.* SANS Institute.

8

Emergency Response

These brave police officers work tirelessly to keep us all safe. They prepare meticulously in case of attacks and put their lives on the line in order to protect the city. They ran to confront danger while encouraging others to run to safety.

—Sadiq Khan, Mayor of London, after March 22, 2017, terrorist attack in Westminster

The events of 9/11 and the anthrax attack ushered in a new world of terrorist action. We've moved beyond the "why." We should shift our mindset to "when?" and "what-if." **What-if analysis** is a proactive approach used to prevent or mitigate certain disasters, extreme events—those human- or nature-generated. What-if questions, when properly posed and answered (with results), can reduce the death toll and overall damage caused by natural disasters and man-made destruction. What-if questions prepare us to react and respond to certain contingencies. And respond we must, because there are certain events we simply can't prevent.

DIB SECTOR CONTINGENCY PLANNING

Emergency response planning, or contingency planning, for extreme events has long been standard practice for safety professionals in industrial systems operations. For many years, prudent practices have required consideration of the potential impact of severe natural events including earthquakes, tornadoes, volcanoes, floods, hurricanes, and blizzards. These possibilities have been included in DIB sector industry infrastructure emergency preparedness and disaster response planning. In addition, many DIB sector production and material distribution facilities have considered the potential consequences of man-caused disasters such as operator error and manufacturer's industrial equipment defects. Currently, DIB sector managers and operators must also consider violence in the workplace. Moreover, at present, as this text has pointed out, there is a new focus of concern: the potential effects of intentional acts by domestic (homegrown) or international (foreign) terrorists.

As a result, the security paradigm has not necessarily changed, but instead has been radically adjusted—reasonable, necessary, and sensible accommodation for and mitigation of just about any emergency situation imaginable have been and continue to be made. Because we cannot foresee all future domestic or intentional acts of terrorism, we must be prepared to shift from the proactive to reactive mode on short notice—in some cases, on very short notice. Unfortunately, there is more we can't prevent than we can prevent. In this chapter we present in outline form a reactive mitigation procedure, the template example for a standard DIB Sector Emergency Response or Contingency Plan, dealings with natural and man-caused disasters, which could also be applied in response to acts of terrorism.

THE EMERGENCY RESPONSE PLAN

Note: The following criteria have not been established as anything other than guidelines and are offered not as definitive or official regulations or procedures but rather as informed advice (based on more than 30 years of safety, industrial hygiene, emergency contingency planning, and security experience) insofar as to the subject matter specific to both public and private sectors.

The goals of an Emergency Response Plan (ERP) are to document and understand the steps needed to:

- Rapidly restore material producing, refining, transporting, generating, transmitting, conserving, building, distributing, maintaining, and controlling systems and components after an emergency
- Minimize DIB sector production process equipment damage
- Minimize impact and loss to customers

- Minimize negative impacts on public health and employee safety
- Minimize adverse effects on the environment
- Provide emergency public information concerning customer service
- Provide hazardous chemical information for first responders and other outside agencies

Although we are concerned with the DIB sector in this text, the USEPA developed *Large Water System Emergency Response Outline: Guidance to Assist Community Water Systems in Complying with the Public Health and Bioterrorism Preparedness and Response Act of 2002* (dated July 2003), with minor adjustments, that can be applied as a template for the DIB sector. This template provides guidance and recommendations to aid facilities in the preparation of emergency response plans under the PL 107-188. The template is provided below.

ERPs do not necessarily need to be one document. They may consist of an overview document, individual Emergency Action Procedures, check lists, additions to existing operations manuals, appendices, and so on. There may be separate, more detailed plans for specific incidents. There may be plans that do not include particularly sensitive information and those that do. Existing applicable documents should be referenced in the ERP (e.g., chemical Risk Management Program, contamination response).

ERP Template

I. Introduction

Safe and reliable operation is vital to every industrial operation. Emergency Response Planning (ERP) is an essential part of managing a DIB sector plant, process, or entity. The introduction should identify the requirement to have a documented emergency response plan (EFP), the goal(s) of the plan (e.g., be able to quickly identify an emergency and initiate timely and effective response action, be able to quickly respond and repair damages to minimize system downtime), and how access to the plan is limited. Plans should be numbered for control. Recipients should sign and date a statement that includes their (1) ERP number, (2) agreement not to reproduce the ERP, and (3) they have read the ERP.

II. Emergency Planning Process

A. Planning Partnerships

The planning process should include those parties who will need to help the DIB sector in an emergency situation (e.g., first responders, law enforcement, public health officials, nearby utilities, local emergency planning committees, testing labs, etc.). Partnerships should track from the DIB sector operation up through local, state, regional, and federal agencies, as applicable and appropriate, and could also document compliance with governmental requirements.

B. General Emergency Response Policies, Procedures, Actions, Documents

A short synopsis of the overall emergency management structure, how other industrial emergency response, contingency, and risk-management plans fit into the ERP for DIB sector emergencies, and applicable policies, procedures, action plans, and reference documents should be cited. Policies should include interconnect agreements with adjacent communities and just how the ERP may affect them.

C. Scenarios

Use your Vulnerability Assessment (VA) findings to identify specific emergency action steps required for response, recovery, and remediation for applicable incident types. In Section V of this plan, specific emergency actions procedures addressing each of the incident types should be addressed.

III. Emergency Response Plan—Policies

A. System Specific Information

In an emergency, DIB sector industries need to have basic information for system personnel and external parties such as law enforcement, emergency responders, repair contractors/vendors, the media, and others. The information needs to be clearly formatted and readily accessible so system staff can find and distribute it quickly to those who may be involved in responding to the emergency. Basic information that may be presented in the emergency response plan are the system's ID number, system name, system address or location, directions to the system, population served, number of service connections, system owner, and information about the person in charge of managing the emergency. Distribution maps, detailed plant drawings, site plans, source/storage/production energy locations, and operations manuals may be attached to this plan as appendices or referenced.

1. PWS ID, Owner, Contact Person
2. Population served and service connections
3. System Components
   a) Pipes and constructed conveyances
   b) Physical barriers
   c) Isolation valves
   d) Chemical treatment, storage, and distribution facilities
   e) Electronic, computer, or other automated systems which are utilized by the DIB sector industry
   f) Emergency power generators (onsite and portable)
   g) The use, storage, or handling of various chemicals
   h) The operation and maintenance of such system components

B. Chain-of-Command Chart developed in Coordination with Local Emergency Planning Committee (internal and/or external Emergency Responders, or both)

1. Contact Name
2. Organization and Emergency Response Responsibility
3. Telephone number(s) (hardwire, cell phones, faxes, e-mail)
4. State 24-hour Emergency Communications Center Telephone

C. Communications Procedures: Who, What, When

During most emergencies, it will be necessary to quickly notify a variety of parties both internal and external to the DIB sector entity. Using the Chain-of-Command Chart and all appropriate personnel from the lists below, indicate who activates the plan, the order in which notification occurs, and the members of the Emergency Response Team. All contact information should be available for routine updating and readily available. The following lists are not intended to be all inclusive—they should be adapted to your specific needs.

1. Internal Notification Lists
   a) Operations Dispatch
   b) DIB Sector Plant Manager
   c) DIB Sector Processing Manager
   d) DIB Production/Storage/Distribution Manager
   e) Facility Managers
   f) Chief Energy Engineer
   g) Director of Engineering
   h) Data (IT) Manager
   i) Maintenance Manager
   j) Other
2. Local Notification
   a) Head of local government (i.e., Mayor, City Manager, Chairman of Board, etc.)
   b) Public Safety Officials—Fire, Local Law Enforcement (LLE), Police, EMS, Safety if a malevolent act is suspected LLE should be immediately notified and in turn will notify the FBI, if required. The FBI is the primary agency for investigating sabotage.
   c) Other Government Entities: Health, Schools, Parks, Finance, Electric, etc.
3. External Notification Lists
   a) State Department of Environmental Quality (DEQ)
   b) USEPA/USDOE/DHS
   c) State Police
   d) State Health Department (lab)
   e) Critical customers (special considerations for hospitals, Federal, State and County government centers, etc.).
   f) Service Aid
   g) Mutual Aid
   h) Chemical Information Sharing and Analysis Center (ISAC)/CHEMTREC
   i) Commercial customers not previously notified
4. Public/Media Notification: When and How to Communicate Effective communications is a key element of emergency response, and a media or communications plan is essential to

good communications. Be prepared by organizing basic facts about the crisis and your DIB sector assets. Develop key messages to use with the media that are clear, brief, and accurate. Make sure your messages are carefully planned and have been coordinated with local and state officials. Considerations should be given to establishing protocols for both field and office staff to respectfully defer questions to the utility spokesperson. Be prepared to list geographic boundaries of the affected area, (e.g., west of highway a, east of highway b, north of highway c, and south of highway d) to ensure the public clearly understands the system boundaries.

E. Personnel Safety

This should provide direction as to how operations staff, emergency responders, and the public should respond to a potential toxic chemical release, including facility evacuation, personnel accountability, proper Personnel Protective Equipment as dictated by the Risk Management Program and Process Safety Management Plan, and whether the nearby public should be sheltered in place or evacuated.

F. Equipment

The ERP should identify equipment that can obviate or significantly lessen the impact of terrorist attacks or other intentional actions on the public health and protect the safety and supply of communities and individuals. The DIB sector facility should maintain an updated inventory of current equipment and repair parts for normal maintenance work.

Because of the potential for extensive or catastrophic damage that could result from a malevolent act, additional equipment sources should be identified for the acquisition and installation of equipment and repair parts in excess of normal usage. This should be based on the results of the specific scenarios and critical assets identified in the vulnerability assessment that could be destroyed. For example, numerous pumps, vats, and mixers specifically designed for DIB sector entities could potentially be destroyed. A certain number of "long-lead" procurement equipment should be inventoried and the vendor information for such unique and critical equipment maintained. In addition, mutual aid agreements with other industries, and the equipment available under the agreement, should be addressed. Inventories of current equipment, repair parts, and associated vendors should be indicated.

G. Property Protection

A determination should be made as to what DIB sector producing, processing, distribution operation/facility should be immediately "locked down," specific access control procedures implemented, initial security perimeter established, a possible secondary malevolent event considered. The initial act may be a divisionary act.

H. Training, Exercises, and Drills

Emergency response training is essential. The purpose of the training program is to inform employees or what is expected of them during an emergency situation. The level of training on an ERP directly affects how well a DIB sector facility's employees can respond to an emergency. This may take the form of orientation scenarios, table-top workshops, functional exercises, etc.

I. Assessment

To evaluate the overall ERP's effectiveness and to ensure that procedures and practices developed under the ERP are adequate and are being implemented, DIB sector industry staff should audit the program on a periodic basis.

IV. Emergency Action Procedures (EAPs)

These are detailed procedures used in the event of an operation emergency or malevolent act. EAPs may be applicable across many different emergencies and are typically common core elements of the overall municipality ERP (e.g., responsibilities, notifications lists, security procedures etc.) and can be referenced.

A. Event classification/severity of emergency
B. Responsibilities of Emergency Director
C. Responsibilities of Incident Commander
D. Emergency Operations Center (EOC) activation
E. Division internal communications and reporting

F. External communications and notifications
G. Emergency telephone list (division internal contacts)
H. Emergency telephone list (off-site responders, agencies, state 24-hour emergency phone number, and others to be notified)
I. Mutual Aid Agreements
J. Contact list of available emergency contractor services/equipment
K. Emergency equipment list (including inventory for each facility)
L. Security and access control during emergencies
M. Facility evacuation and lockdown and personnel accountability
N. Treatment and transport of injured personnel (including electrocution and petrochemical exposure)
O. Petrochemical records—to compare against historical results for base line
P. List of available labs for emergency use
Q. Emergency sampling and analysis (petrochemical)
R. Water use restrictions during emergencies
S. Alternate temporary chemical supplies during emergencies
T. Isolation plans for chemical supply, treatment, storage, and distribution systems
U. Mitigation plans for neutralizing, flushing, and collecting spilled chemicals
V. Protection of vital records during emergencies
W. Record keeping and reporting (FEMA, DHS, DOT, OSHA, EPA, and other requirements) (it is important to maintain accurate financial records of expenses associated with the emergency event for possible federal reimbursement)
X. Emergency program training, drills/and tabletop exercises
Y. Assessment of emergency management plan and procedures
Z. Crime scene preservation training and plans
AA. Communication Plans:
    1. Police
    2. Fire
    3. Local Government
    4. Media
    5. And so forth

BB. Administration and logistics, including EOC, when established
CC. Equipment needs/maintenance of equipment
DD. Recovery and restoration of operations
EE. Emergency event closeout and recovery

V. Incident-Specific Emergency Action Procedures (EAPs)

Incident-Specific EAPs are action procedures that identify specific steps in responding to an operational emergency of malevolent act.

A. General Response to Terrorist Threats (Other than Bomb Threat and Incident-Specific Threats)
B. Incident-Specific Response to Man-Made or Technological Emergencies
   1. Contamination event (Articulated Threat with Unspecified Materials)
   2. Contamination Threat at a Major Event
   3. Notification from Health Officials of Potential Contamination
   4. Intrusion through Supervisory Control and Data Acquisition (SCADA)
C. Significant structural damage resulting from intentional act
D. Customer complaints
E. Severe weather response (snow, ice, temperature, lightning)
F. Flood response
G. Hurricane and/or tornado response
H. Fire response
I. Explosion response
J. Major vehicle accident response
K. Electrical power outage response
L. Water supply interruption response
M. Transportation accident response—barge, plane, train, semi-trailer/tanker
N. Contaminated/tampered with water treatment chemicals
O. Earthquakes response
P. Disgruntled employees response (i.e., workplace violence)
Q. Vandals response
R. Bomb threat response
S. Civil disturbance/riot/strike
T. Armed intruder response

U. Suspicious mail handling and reporting
V. Hazardous chemical spill release response (including Material Safety Data Sheets)
W. Cybersecurity/Supervisory Control and Data Acquisition (SCADA) system attack response (other than incident specific, e.g., hacker)

VI. Next Steps
  A. Plan Review and Approval
  B. Practice and Plan to Update (as necessary, once every year recommended)
    1. Training requirements
    2. Who is responsible for conducting training, exercises, and emergency drills
    3. Update and assessment requirements
    4. Incident-specific requirements

VII. Annexes
  A. Facility and Location Information
    1. Facility maps
    2. Facility drawings
    3. Facility descriptions/layout

VIII. References and Links
  A. Department of Homeland Security— https://www.dhs.gov
  B. Environmental Protection Agency—https://www.epa.gov
  C. The American Water Works Association (AWWA)—https://www.awwa.org
  D. The Center for Disease Control and Prevention— https://www.cdc.gov
  E. Federal Emergency management Agency—https://www.fema.gov

OSHA AND EMERGENCY RESPONSE PLANNING

Even though no single OSHA standard is dedicated specifically to the issue of planning for DIB sector facility/site emergencies, all OSHA standards are written for the purpose of promoting a safe, healthy, accident-free, and hence emergency-free workplace. Therefore, OSHA standards do play a role in emergency prevention.

OSHA's standards, therefore, should be considered when developing emergency plans. A first step when developing emergency response plans is to review the guidelines presented earlier and to apply any and all applicable OSHA standards to your emergency response plan. This review of applicable guidelines and pertinent OSHA standards can help the managers of DIB sector industrial facilities identify and then correct conditions that might exacerbate emergency situations before they occur.

Typically, when we think of emergency response plans for the workplace, we often imagine the obvious. For example, the first workplace emergency that might come to mind is fire—a major concern because fire in the workplace is something that can happen, that happens more often than we might think, and because fire can be particularly devastating—in ways we know all too well. Most employees do not need to be informed about the dangers of fire. However, employers have the responsibility to do just this—to inform and train employees on fire, fire prevention, and fire protection. Many local codes go beyond this information requirement, insisting that employers develop and implement a fire emergency response and/or evacuation plan. The primary emphasis has been on the latter—evacuation. However, if the employer equips a workplace with fire extinguishers and other fire-fighting equipment, and expects its employees to respond aggressively to extinguish workplace fires, then not only must the facility have an emergency response plan, the employer must also ensure that all company personnel called upon to fight the fire are completely trained on how to do so safely; this is stipulated in OSHA's 29 CFR 1910.156(c).157(g).

Another commonly considered workplace emergency response plan is designed and implemented for medical emergencies. Many DIB facilities satisfy this requirement simply by directing their employees to call 911 or some other emergency number whenever a medical emergency occurs in the workplace. Other facilities, though, may require employees to provide emergency first aid. When the employer chooses the employee-supplied first aid option, certain requirements must be met before any employee can legally administer first aid. First, the first aid responder must be trained and certified to administer first aid. This training aspect must also include training on OSHA's Bloodborne Pathogen Standard. This standard requires that the employee is trained on the dangers inherent with handling and being exposed to human body fluids. The employee must also be trained on how to protect him or herself from body fluid contamination. If the first aid responder or anyone else is exposed to and contaminated by body fluids, the employer must make available the hepatitis B vaccine and vaccination series to all employees who have occupational exposure, and post-exposure evaluation and follow-up to all employees who have had an exposure incident; this requirement is stipulated in OSHA's 29 CFR 1910.1030.

A third type of emergency response plan required for implementation in selected (covered) facilities is OSHA's 29 CFR 1910.120 (Hazardous Waste Operations and Emergency Response—HAZWOPER)—for releases of hazardous chemical materials. Unless the facility operator can demonstrate that the operation does not involve employee exposure or the reasonable possibility for employee exposure to safety or health hazards, the following operations are covered:

1. Clean-up operations required by a governmental body involving hazardous substances conducted at uncontrolled hazardous waste sites, state priority site lists, sites recommended by the EPA, NPL, and initial investigations of government identified sites that are conducted before the presence or absence of hazardous substance has been ascertained
2. Corrective actions involving clean-up operations at sites covered by the Resource Conservation and Recovery Act of 1976 (RCRA)
3. Voluntary clean-up operations at sites recognized by federal, state, local or other governmental bodies as uncontrolled hazardous waste sites
4. Operations involving hazardous petrochemical waste conducted at treatment, storage, disposal (TSD) facilities regulated by RCRA
5. Emergency response operations for releases of, or substantial threats of releases of, hazardous petrochemical substances without regard to the location of the hazard

The final requirement impacts the largest number of facilities that meet the criteria requiring full compliance with 29 CFR 1910.120 HAZWOPER, because many such facilities do not normally handle, store, treat, or dispose of hazardous chemical waste, but do use or produce hazardous materials in their processes.

Because the use of hazardous petrochemical materials could lead to an emergency from the release or spill of such materials, facilities using these materials must develop and employ an effective site emergency response plan.

Before we discuss the basic goals of an effective emergency response plan from an OSHA compliance point of view, we should define "emergency response." Considering that individual facilities are different, with different dangers and different needs, defining emergency response is not always easy. However, for our purposes, we use the definition provided by CoVan (1995, 54):

> Emergency response is defined as a limited response to abnormal conditions expected to result in unacceptable risk requiring rapid corrective action to prevent harm to personnel, property, or system function.

CoVan makes another important point about emergency response, one critical for the site manager and/or safety professional. He points out that "although emergency response and engineering tends toward prevention, emergency response is a skill area that safety engineers must be familiar with both because of regulations and good engineering practice." "Good Engineering Practice" is the law by which all competent managers and safety professionals work and live.

Now that we have defined emergency response, let's move on to the basic goals of an effective emergency response plan from an OSHA compliance standpoint. The goals are twofold:

1. Minimize injury to facility personnel.
2. Minimize damage to the facility, and then return to normal operation as soon as possible.

Obviously, these goals make a great deal of good sense. However, you may be wondering about the language used—a couple of key words: "facility personnel" and "damage to the facility." Remember that we are talking about OSHA requirements here. Under OSHA the primary emphasis is protecting the worker—protecting the worker's health and safety is OSHA's only focus.

DID YOU KNOW?

OSHA's 29 CFR 1910.151 (a), (b), and (c), Medical Services and First Aid standard, requires employers to

1. have readily available medical personnel for advice and consultation on matters of health
2. in the absence of an infirmary, clinic, or hospital in near proximity to the workplace, to train a person or persons to render first aid and to have adequate first aid supplies readily available
3. where the eyes or body of any person may be exposed to injurious corrosive materials, to provide suitable facilities within the work area for quick drenching of or flushing of the eyes and body.

In auditing various industrial workplaces, we found that this particular OSHA standard is one of the most misunderstood by employers. Based on our experience, we found that OSHA's use of the words "near proximity to" and "adequate" contribute to the vagary and ambiguity of this standard. Most employers understand the need to provide first aid supplies in the workplace, but generally think that this is the extent of their responsibilities in this regard. Moreover, many workplaces do train their personnel on basic first aid and CPR, but do not require (in written job descriptions) that the trained employees respond to a workplace medical emergency.

We found that employers are reluctant to have employees respond to workplace medical emergencies because most employees do not want to respond. They do not want to be exposed to gore and the victim's pain. They also do not want to assume liability for trying to aid an injured victim.

- What about people who live offsite—the site's neighbors?
- What about the environment?

These questions make the point we emphasize here. Again, OSHA is not normally concerned about the environment, unless contamination of the environment (at the worksite) might adversely impact the worker's safety and health. The neighbors? Again, OSHA's focus is the worker. One OSHA compliance officer explained to us that if the employer takes every necessary step to protect its employees from harm involving the use or production of hazardous chemical materials, then the surrounding community should have little to fear.

This statement is puzzling to us. We asked the same OSHA compliance officer about those incidents beyond the control of the employer—about accidents that could not only put employees in harm's way, but also endanger the surrounding community. The answer? "Well, that's the EPA's bag—we only worry about the worksite and the worker."

Fortunately, OSHA, in combination with the U.S. Environmental Protection Agency (USEPA) has taken steps to overcome this blatant shortcoming (we like to think of it as an oversight). Under OSHA's Process Safety Management (PSM) and EPA's Risk Management Planning (RMP) directives, chemical spills and other chemical accidents that could impact both the environment and the "neighbors" have now been properly addressed. What PSM and RMP really accomplish is changing the typical twofold goal of an effective emergency response plan to a threefold goal.

Let's point out that the accomplishment of these two- or threefold goals or objectives is essential in emergency response. Accomplishing these goals or objectives requires an extensive planning effort prior to the emergency ("prior" being the keyword, because the attempts to develop an emergency response plan when a disaster is occurring or after one has occurred is both futile and stupid). The site manager and/or safety professional must never forget that while hazards in any facility can be reduced, risk is an element of everyday existence, and therefore cannot be totally eliminated. The manager/safety professional's goal must be to keep risk to an absolute minimum. Again, to accomplish this, planning is critical—as well as essential, and should be accomplished well in advance of emergency, because program testing should be part of the overall program design.

We pointed out earlier that most emergency plans address fire, medical emergencies, and the accidental release or spills of petrochemical materials. Note that the development of emergency response plans should also factor in other possible emergencies—natural disasters, floods, explosions, and/or weather-related events that could occur, and certainly will occur. Now, emergency response to terrorist activity or threats must also be added to the list.

Site emergency response plans should include:

- Assessment of risk
- Chain of command for dealing with emergencies
- Assessment of resources
- Training
- Incident command procedures
- Site security
- Public relations

The Federal Emergency Management Agency (better known as FEMA), the U.S. Army Corps of Engineers, and several other agencies, as well as numerous publications, provide guidance on how to develop a site emergency response plan. Local agencies (such as fire departments, emergency planning commissions/agencies, HazMat teams, and Local Emergency Planning Committees [LEPCs]) also provide information on how to design a site plan. All of these agencies typically recommend that a site's plan contain the elements listed in table 8.1.

In the site manager/safety professional's effort to incorporate and manage a facility emergency response plan, and in the response itself, two elements mentioned earlier (security considerations and public relations) must be given special attention. If not handled correctly, the lack of effective security measures and/or improper public relations can turn an already disastrous incident into a mega-disaster.

In planning security considerations, provision should be made to have a well-trained security team limit site access to only those people and that equipment that will assist in coping with and resolving the emergency.

Public relations (PR) can be a tricky enterprise. The person identified to interface with the media must have thorough knowledge of the site, process, and personnel involved. The PR person must also have access to the highest

Table 8.1. Site Emergency Response Plan

Emergency Response Notification	List of who to call and information to pass on when an emergency occurs
Record of Changes	Table of changes and dates for them
Table of Contents/ Introduction	The purpose, objective, scope, applications, policies and assumptions for the plan
Emergency Response Operations	Details what actions must take place
Emergency Assistance Telephone Numbers	A current list of people and agencies who may be needed in an emergency
Legal Authority and Responsibility	References the laws and regulations that provide the authority for the plan
Chain of Command	Response organization structure and responsibilities
Disaster Assistance and Coordination	Where additional assistance may be obtained when the regular response organizations are over-burdened
Procedures for Changing or Updating the Plan	Details who makes changes and how they are made and implemented
Plan Distribution	List of organizations and individuals who have been give a copy of the plan
Spill Cleanup Techniques	Detailed information about how response teams should handle cleanups
Cleanup/Disposal Resources	List of what is available, where it is obtained, and how much is available
Consultant Resources	List of special facilities and personnel who may be valuable in a response
Technical Library/References	List of libraries and other information sources that may be valuable for those preparing, updating, or implementing the plan
Hazards Analysis	Details the kinds of emergencies that may be encountered, where they are likely to occur, what areas of the community may be affected and the probability of occurrence
Documentation of Spill Events	The various incident and investigative reports on spills that have occurred
Hazardous Materials Information	Listing of hazardous materials, their properties, response data, and related information
Dry Runs	Training exercises for testing the adequacy of the plan, training personnel, and introducing changes

Planning Guide and Checklist for Hazardous Materials Contingency Plans, FEMA-10, Federal Emergency Management Agency, Washington, D.C., July 1981. Adaptation from Brauer, R. L., *Safety and Health for Engineers*. New York: Van Nostrand Reinhold, 1994.

levels of site management. Otherwise, he or she will not be able to deal with the public/media effectively.

THE BOTTOM LINE

Because industrial emergencies (in less than extreme conditions) can seriously affect the surrounding community and environment, and because poor planning and/or panic can only make a bad situation worse, and can also lead to additional injury and death, your role as DIB sector site manager or site safety professional in emergency response is doubly important. A crisis out of hand can easily devastate a community—and your organization is (or should be) an active member of your community. By ensuring less than effective emergency response, DIB site managers endanger not only themselves and their organizations but also endanger their organization's community and standing as well.

REFERENCES AND RECOMMENDED READING

Brauer, R. L. (1994). *Safety and Health for Engineers.* New York: Van Nostrand Reinhold.

CoVan, J. (1995). *Safety Engineering.* New York: John Wiley and Sons.

Office of the Federal Register, 29 CFR 1910.120. Washington, DC: Office of the Federal Register, 1987.

Planning Guide and Checklist for Hazardous Materials Contingency Plans. FEMA-10, Federal Emergency Management Agency, Washington, DC, July 1981.

Safety and Health Requirements Manual, rev. ed. EM 385-1-1, U.S. Army Corps of Engineers, Washington, DC, October 1987.

Spellman, F. R. (1997). *A Guide to Compliance for Process Safety Management Planning (PSM/RMP).* Lancaster, PA: Technomic Publishing Company.

USDOE (2008). *Emergency Support Function #12—Energy Annex.* Washington, DC: U.S. Department of Energy.

USDOE (2010. *Energy) Sector-Specific Plan: An Annex to the National Infrastructure Protection Plan.* Washington, DC: USDOE.

USEPA (2002). *Water Utility Response, Recovery & Remediation Guidance for Man-made and/or Technological Emergencies.* Washington, DC: United States Environmental Protection Agency.

USEPA (2003). *Large Water System Emergency Response Plan Outline: Guidance to Assist Community Water Systems in Complying with the Public Health Security and Bioterrorism Preparedness and Response Act of 2002.* EPA 810-F-03-007 accessed 6/06 at www.epa.gov/safewater/security. Washington, DC: United States Environmental Protection Agency.

9

Security Techniques and Hardware

> The kind of challenge that we face from terrorism and other extremists today is more like a public health problem or crime that we are going to have to battle on an enduring basis, and there won't be a moment when any of these groups say we surrender or we quit.
>
> —Robert Gates, 2016

THE MULTI-BARRIER APPROACH

Ideally, all DIB sector sites/facilities would be secured in a layered fashion. Layered security (aka multi-barrier) systems are vital. Using the "protection-in-depth" principle requiring that an adversary defeat several protective barriers or security layers to accomplish its goal, the DIB sector infrastructure can be made more secure. "Protection-in-depth" is a term commonly used by the military to describe security measures that reinforce one another, masking the defense mechanisms from view of intruders and allowing the defender time to respond to intrusion or attack.

A prime example of the use of the multi-barrier approach to ensure security and safety is demonstrated by the practices of the bottled water industry. In the aftermath of 9/11 and the increased emphasis on Homeland Security, a shifted paradigm of national security and vulnerability awareness emerged. Recall that immediately after the 9/11 tragedies, emergency responders and others responded quickly and worked to exhaustion. In addition to the emergency responders, bottled water companies responded immediately by donating several million bottles of water to the crews at the crash sites in New York, at the Pentagon, and in Pennsylvania. International Bottled Water Association (IBWA 2004) reported that "within hours of the first attack, bottled water was delivered where it mattered most; to emergency personnel on the scene who required ample water to stay hydrated as they worked to rescue victims and clean up debris."

Bottled water companies continued to provide bottled water to responders and rescuers at the 9/11 sites throughout the post-event(s) process(es). These patriotic actions by the bottled water companies, however, beg the question: How do we ensure the safety and security of the bottled water provided to anyone? IBWA has the answer: using a multi-barrier approach, along with other principles, will enhance the safety and security of bottled water. IBWA (2004) describes its multi-barrier approach as follows:

> **A multi-barrier approach**—Bottled water products are produced utilizing a multi-barrier approach, from source to finished product, that helps prevent possible harmful contaminants (physical, chemical or microbiological) from adulterating the finished product as well as storage, production, and transportation equipment. Measures in a multi-barrier approach may include source protection, source monitoring, reverse osmosis, distillation, filtration, ozonation or ultraviolet (UV) light. Many of the steps in a multi-barrier system may be effective in safeguarding bottled water from microbiological and other contamination. Piping in and out of plants, as well as storage silos and water tankers are also protected and maintained through sanitation procedures. In addition, bottled water products are bottled in a controlled, sanitary environment to prevent contamination during the filling operation. (3)

So, within DIB sector infrastructure security, let's consider a protection-in-depth strategy: using several forms of security techniques and/or devices against an intruder that thus does not rely on one single defensive mechanism to protect infrastructure. By implementing multiple layers of security, a hole or flaw in one layer is covered by the other layers. An intruder will have to intrude through each layer without being detected in the process—the layered approach implies that no matter how an intruder attempts to accomplish his or her goal, he or she will encounter effective elements of the physical protection system.

For example, look at figure 9.1: an effective security layering approach requires that an adversary penetrate multiple, separate barriers to gain entry to a critical TARGET at a DIB facility. As shown in figure 9.1, protection in depth (multiple layers of security) helps to ensure that the security system remains effective in the event of a failure or an intruder bypassing a single layer of security.

As shown in figure 9.1, layered security starts with the outer perimeter (the fence—the first line of physical security) of the facility and goes inward to the facility, the buildings, structures, other individual assets, and finally to the contents of those buildings—the TARGETs.

FIGURE 9.1
Layered, Multiple-Barrier Approach to Security

The area between the outer perimeter and structures or buildings is known as the **site**. This open site area provides an incomparable opportunity for early identification of an unauthorized intruder and initiation of early warning/response. This open space area is commonly used to calculate the **standoff distance**; that is, it is the distance between the outside perimeter (public areas to the fence) to TARGET or critical assets (buildings/structures) inside the perimeter (inside the fence line—the restricted access area).

The open area, between perimeter fence and TARGET (e.g., operations center), if properly outfitted with various security devices, can also provide layered protection against intruders. For example, lighting is a deterrent. Based on personal experience, I think an open area within the facility site that is almost as well lighted at night as would be expected during daylight hours should be the rule of thumb. In addition, strategically placed motion detectors along with crash barriers at perimeter gate openings and in front of vital structures are also recommended. Armed, mobile guards who roam the interior of the site on a regular basis provide the ultimate in site-area security.

The next layer of physical security is the outside wall of the target structure(s) itself. Notwithstanding door, window, and/or skylight entry, walls prevent most intruders from easy entry. If doors can only be entered using card reader access, security is shored up or enhanced to an extent. The same can be said for windows and skylights that are fashioned small enough to prohibit normal human entry. These same "weak" spots in buildings can be bastioned with break-proof or reinforced security glass.

The final layer of security is provided by properly designed interior features of buildings. Examples of these types of features include internal doors and walls, equipment cages, and backup or redundant equipment.

The conditions I've described above are "perfect world" conditions; that is, those conditions that we "would want" (i.e., the security manager's proverbial wish list) to be incorporated into the design and installation of new DIB sector infrastructure. Post 9/11, in a not so perfect world, many of the peripheral (fence line) measures described above are more difficult to incorporate into DIB infrastructure. This is not to say that DIB sector sites and facilities do not have fence lines or fences—many of them do. These fences are designed to keep vandals, thieves, and trespassers out. But many of these facilities were constructed several years ago before urban encroachment literally encircled them, allowing, at present, little room for security stand backs or setbacks to be incorporated into electrical power stations, plants, and/or critical equipment locations. When you sit down to plan a security upgrade, you have to take account these kind of "real-world" factors.

Responsible managers of DIB sector infrastructure have four primary security areas to manage:

- **Physical Security**—in the DIB sector, physical security techniques and practices are the most effective at sites where a systems approach is in place; that is, where detection, assessment, communication, and response are planned and supported by resources, procedures, and policies.
- **Cyber/Information Technology Security**—only the use of SCADA and other key operating systems that have been properly vetted and scrubbed of alleged Chinese and/or Russian Trojan horses hacked into the North American electrical grid is important. The only positive way to ensure the security of the North American grid is to disconnect its cyber and other digital systems from the Internet. This step is impractical at the present time, but points to the need to conduct frequent audits of the system and install firewall protection in SCADA and other systems to prevent hacking. Frequent third-party penetration testing is advised.
- **Employment Screening**—helps to mitigate the threat from the "enemy at the water cooler" (inside the organization). I am always amazed whenever I conduct security audits for various companies. Often, a simple check such as reviewing an employee's driving record often reveals that the employee has no license, is driving on a suspended license, or has a horrific driving record. Hiring standards and pre-employment background investigations may help ensure the trustworthiness and reliability of personnel who have unescorted access to critical facilities.
- **Protecting Potentially Sensitive Information**—the old saying that goes a secret can best be held between three people so long as two of the three are dead makes the point that reducing the likelihood that information could be used by those intent on disrupting operations or causing destruction at DIB sector plants/sites is crucial. Information should only be shared within an organization on a need-to-know basis.

For existing facilities, security upgrades should be based on the results generated from the vulnerability assessment, which characterizes and prioritizes those assets that may be targeted. Those vulnerabilities identified must be protected.

In the following sections, I describe various security hardware and/or devices. These devices serve the main purpose of providing security against physical and/or digital intrusion. That is, they are designed to delay and deny intrusion and are normally coupled with detection and assessment technology. Possible additional security measures, based on the vulnerability assessment (see chapter 5) that may be recommended include the following (NAERC 2002):

- Electronic security
- Closing non-essential perimeter and internal portals
- Physical barriers such as bollards or highway (Jersey) walls
- Fencing
- Lighting
- Security surveys
- Vulnerability assessments
- Availability of security resources
- General personnel and security officer training
- Law enforcement liaison
- Ensure availability of essential spare parts (machines, wire, pipe, valves, transformers, etc.) for critical facilities

Remember: no matter the type of security device or system employed, DIB sector systems cannot be made immune to all possible intrusions or attacks. Whenever a facility safety/security manager tells me that he or she has secured their site 100 percent, I am reminded of Schneier's (2000) view of security: "You can't defend. You can't prevent. The only thing you can do is detect and respond." Simply, when it comes to making "anything" absolutely secure from intrusion or attack, there is inherently, or otherwise, no silver bullet.

Before I describe security hardware devices, keep in mind that in addition to these devices there are a few employee practices and actions that can be taken to protect DIB assets. For example, when a computer system fails and must be disposed of, how is it disposed of? Is there a procedure or practice in place to prevent the valuable information on the system's hard drive from being pulled from a trash heap or from a dumpster dive and used by potential enemies? Are shredders used? Are they state-of-the-art shredders that prevent scraps from being reassembled by enemies? Are building cleaning crews properly vetted and supervised? Do you have a team that routinely inspects suspended ceilings for bugs, cameras, and listening recorders? Do you have key stroke reader capability; that is, can you record what messages are being sent by employees? Do you routinely check hard-wired phone lines? Have your removed all door signs that tell anyone what is on the other side of the room? Have you trained your employees to be slightly suspicious of just about anything and everything? Have you opened your manholes lately to see what is inside?

These practices do not require security hardware such as barriers, motion detectors, fences, locks, biometric systems, video cameras, armed guards, electrified fences, and so forth. What they require instead is common sense, awareness, and alert and engaged supervisors and employees. The point is what is heard at the water cooler is sometimes more significant than any security alarm apparatus can provide.

SECURITY HARDWARE DEVICES[1]

Infrastructure security devices or products described below into four general categories:

- Physical asset monitoring and control devices
- Cyber protection devices
- Communication/integration
- Environmental monitoring devices

Physical Asset Monitoring and Control Devices

Aboveground, Outdoor Equipment Enclosures

DIB sector facilities and sites usually consist of multiple components spread over a wide area and typically include a centralized production and distribution center, as well component storage facilities that are typically distributed at multiple locations throughout the area. One of the primary reasons for constructing structural components that house operational equipment above-

1. Please note that the following USEPA (2005) security asset and device recommendations were first made for the water/wastewater critical infrastructure, but these recommendations are applicable to all other critical infrastructure sectors—including the DIB sector.

ground eliminates the safety risks associated with confined space entry, which is often required for the maintenance of equipment located below ground. In addition, space restrictions often limit the amount of equipment that can be located inside, and there are concerns that some types of equipment (such as backflow prevention devices—to prevent chemicals and fuel wastes from entering plant and offsite potable water systems) can, under certain circumstances, discharge fuel slurry or waste mixtures that could flood pits, vaults, or equipment rooms. In regard to electrical power, electrical substations are not usually suited for underground installation. Therefore, many pieces of critical electrical manufacturing equipment are located outdoors and aboveground in configurations that are properly fenced, insulated, or isolated to prevent accidental electrical shock or short circuits/fires in equipment.

Experience demonstrates that many different system components can be and often are installed outdoors and aboveground. Examples of these types of components could include:

- Backflow prevention devices
- Air release and control valves
- Pressure vacuum breakers
- Oil and gas pumps and motors
- Petrochemical storage and feed equipment
- Meters
- Sampling equipment
- Instrumentation
- Electrical substations
- Oil and natural gas pipelines

One of the most effective security measures for protecting aboveground equipment, where feasible, is to place it inside a building or exterior fenced structure. When/where this is not possible, enclosing the equipment or parts of the equipment using some sort of commercial or homemade add-on structure may help to prevent tampering with the equipment. These types of add-on structures or enclosures, which are designed to protect people and animals from electrocution and to protect equipment both from the elements and from unauthorized access or tampering, typically consist of a box-like fenced structure that is placed over or around the entire component, or over/around critical parts of the component (valves, etc.), and is then secured to delay or prevent intruders from tampering with the equipment. The enclosures are typically locked or otherwise anchored to a solid foundation, which makes it difficult for unauthorized personnel to remove the enclosure and access the equipment.

Standardized aboveground enclosures are available in a wide variety of materials, sizes, and configurations. Many options and security features are also available for each type of enclosure, and this allows system operators the flexibility to customize an enclosure for a specific application and/or price range. In addition, most manufacturers can custom-design enclosures if standard, off-the-shelf enclosures do not meet a user's needs.

Many of these enclosures are designed to meet certain standards. For example, the American Society of Sanitary Engineers (ASSE) has developed Standard #1060, *Performance Requirements for Outdoor Enclosures for Backflow Prevention Assemblies*. If an enclosure will be used to house backflow preventer, this standard specifies the acceptable construction materials for the enclosure, as well as the performance requirements that the enclosure should meet, including specifications for freeze protection, drainage, air inlets, access for maintenance, and hinge requirements. ASSE #1060 also states that the enclosure should be lockable to enhance security.

Electrical substation and electrical equipment enclosures must meet the requirements and recommendations of various OHSA Standards, National Fire Protection Association (NFPA), National Electrical Codes (NEC), Institute of Electrical and Electronic Engineers (IEEE), and local code requirements.

Equipment enclosures can generally be categorized into one of four main configurations, which include:

- One piece, drop-over enclosures
- Hinged or removable top enclosures
- Sectional enclosures
- Shelters with access locks

All enclosures, including those with integral floors, must be secured to a foundation to prevent them from being moved or removed. Unanchored or poorly anchored enclosures may be blown off the equipment being protected, or may be defeated by intruders. In either case, this may result in the equipment beneath the enclosure becoming exposed and damaged. Therefore, ensuring that the enclosure is securely anchored will increase the security of the protected equipment.

The three basic types of foundations that can be used to anchor the aboveground equipment enclosure are concrete footers, concrete slabs-on-grade, or manufactured fiberglass pads. The most common types of foundations utilized for equipment enclosures are standard or slab-on-grade footers; however, local climate and soil conditions may dictate whether either of these types of foundations can be used. These foundations can be either precast or poured in place at the installation site. Once the foundation is installed and properly cured, the equipment enclosure is bolted or anchored to the foundation to secure it in place.

An alternative foundation, specifically for use with smaller hot box enclosures, is a manufactured fiberglass pad known as the Glass Pad™. The Glass Pad™ has the center cut out so that it can be dropped directly over the piece of equipment being enclosed. Once the pad is set level on the ground, it is backfilled over a two-inch flange located around its base. The enclosure is then placed on top of the foundation, and is locked in place with either a staple- or a slotted-anchor, depending on the enclosure configuration.

One of the primary attributes of a security enclosure is its strength and resistance to breaking and penetration. Accordingly, the materials from which the enclosure is constructed will be important in determining the strength of the enclosure, and thus its usefulness for security applications. Enclosures are typically manufactured for either fiberglass or aluminum. With the exception of the one piece, drop-over enclosure, which is typically fabricated from fiberglass, each configuration described above can be constructed from either material. In addition, enclosures can be custom-manufactured from polyurethane, galvanized steel, or stainless steel. Galvanized or stainless steel is often offered as an exterior layer, or "skin," for an aluminum enclosure. Although they are typically utilized in underground applications, precast concrete structures can also be used as aboveground equipment enclosures. However, precast structures are much heavier and more difficult to maneuver than are their fiberglass and aluminum counterparts. Concrete is also brittle, and that can be a security concern, however, products can be applied to concrete structures to add strength and minimize security risks (i.e., epoxy coating). Because precast concrete structures can be purchased from any concrete producers, this text does not identify specific vendors for these types of products.

In addition to the construction materials, enclosure walls can be configured or reinforced to give them added strength. Adding insulation is one option that can strengthen the structural characteristics of an enclosure; however, some manufacturers offer additional features to add strength to exterior walls. For example, while most enclosures are fabricated with a flat wall construction, some vendors manufacture fiberglass shelters with ribbed exterior walls. These ribs increase the structural integrity of the wall and allow the fabrication of standard shelters up to twenty feet in length. Another vendor has developed a proprietary process that uses a series of integrated fiberglass beams that are placed throughout a foam inner core to tie together the interior and exterior walls and roof. Yet another vendor constructs aluminum enclosures with horizontal and vertical redwood beams for structural support.

Other security features that can be implemented on aboveground, outdoor equipment enclosures include locks, mounting brackets, tamper-resistant doors, and exterior lighting.

Active Security Barriers (Crash Barriers)

Active security barriers (also known as crash barriers) are large structures that are placed in roadways at entrance and exit points to protected facilities to control vehicle access to these areas. These barriers are placed perpendicular to traffic to block the roadway, so that the only way that traffic can pass the barrier is for the barrier to be moved out of the roadway. These types of barriers are typically constructed from sturdy materials, such as concrete or steel. They are also designed at a certain height off the roadway so that vehicles cannot go over them.

The key difference between active security barriers, which include wedges, crash beams, gates, retractable bollards, and portable barricades; and passive security barriers, which include non-moveable bollards, Jersey barriers, and planters, is that active security barriers are designed so that they can be raised and lowered or moved out of the roadway easily to allow authorized vehicles to pass them. Many of these types of barriers are designed so that they can be opened and closed automatically (i.e., mechanized gates, hydraulic wedge barriers), while others are easy to open and close manually (swing crash beams, manual gates). In contrast to active barriers, passive barriers are permanent, non-movable barriers, and thus they are typically used to protect the perimeter of a protected facility, such as sidewalks and other areas that do not require vehicular traffic to pass them. Several of the major types of active security barriers such as wedge barriers, crash beams, gates, bollards, and portable/removable barricades are described below.

Wedge barriers are plated, rectangular steel buttresses approximately 2–3 feet high that can be raised and lowered from the roadway. When they are in the open position, they are flush with the roadway and vehicles can pass over them. However, when they are in the closed (armed) position, they project up from the road at a 45-degree angle, with the upper end pointing toward the oncoming vehicle and the base of the barrier away from the vehicle. Generally, wedge barriers are constructed from heavy gauge steel, or concrete that contains an impact-dampening iron rebar core that is strong and resistant to breaking or cracking, thereby allowing them to withstand the impact from a vehicle attempting to crash through them. In addition, both of these materials help to transfer the energy of the impact over the barrier's entire volume, thus helping to prevent the barrier from being sheared off its base. In addition, because the barrier is angled away from traffic, the force of any vehicle impacting the barrier is distributed over the entire surface of the barrier and is not concentrated at the base, which helps prevent the barrier from breaking off at the base. Finally, the angle of the barrier helps hang up any vehicles attempting to drive over it.

Wedge barriers can be fixed or portable. Fixed wedge barriers can be mounted on the surface of the roadway ("surface-mounted wedges") or in a shallow mount in the road's surface, or they can be installed completely below the road surface. Surface-mounted wedge barricades operate by rising from a flat position on the surface of the roadway, while shallow-mount wedge barriers rise from their resting position just below the road surface. In contrast, below-surface wedge barriers operate by rising from beneath the road surface. Both the shallow-mounted and surface-mounted barriers require little or no excavation, and thus do not interfere with buried utilities. All three barrier mounting types project above the road surface and block traffic when they are raised into the armed position. Once they are disarmed and lowered, they are flush with the road; thereby allowing traffic to pass.

Table 9.1. Pros and Cons of Wedge Barriers

Pros	*Cons*
Can be surface-mounted or completely installed below the roadway surface.	Installations below the surface of the roadway will require construction that may interfere with buried utilities.
Wedge barriers have a quick response time (normally 3.5–10.5 seconds, but barrier can be 1–3 seconds in emergency situations). Because emergency activation of the barrier causes more wear and tear on the system than does normal activation, it is recommended for use only in true emergency situations.	Regular maintenance is needed to keep wedge fully operational.
Surface or shallow-mount wedge barricades can be utilized in locations with a high water table and/or corrosive soils.	Improper use of the system may result in authorized vehicles being hung up by the barrier and damaged. Guards must be trained to use the system properly to ensure that this does not happen. Safety technologies may also be installed to reduce the risk of the wedge activating under an authorized vehicle.
All three wedge barrier designs have a high crash rating, thereby allowing them to be employed for higher security applications.	
These types of barrier are extremely visible, which may deter potential intruders.	

USEPA (2005).

Installing rising wedge barriers requires preparation of the road surface. Installing surface-mounted wedges does not require that the road be excavated; however, the road surface must be intact and strong enough to allow the bolts anchoring the wedge to the road surface to attach properly. Shallow-mount and below-surface wedge barricades require excavation of a pit that is large enough to accommodate the wedge structure, as well as any arming/disarming mechanisms. Generally, the bottom of the excavation pit is lined with gravel to allow for drainage. Areas not sheltered from rain or surface runoff can install a gravity drain or self priming pump. Table 9.1 lists the pros and cons of wedge barriers.

Crash beam barriers consist of aluminum beams that can be opened or closed across the roadway. While there are several different crash beam designs, every crash beam system consists of an aluminum beam that is supported on each side of the made by a solid footing or buttress, which is typically constructed from concrete, steel, or some other strong material. Beams typically contain an interior steel cable (typically at least one inch in diameter) to give the beam added strength and rigidity. The beam is connected by a heavy duty hinge or other mechanism to one of the footings so that it can swing or rotate out of the roadway when it is open, and can swing back across the road when it is in the closed (armed) position, blocking the road and inhibiting access by unauthorized vehicles. The non-hinged end of the beam can be locked into its footing, thus providing anchoring for the beam on both sides of the road and increasing the beam's resistance to any vehicles attempting to penetrate through it. In addition, if the crash beam is hit by a vehicle, the aluminum beam transfers the impact energy to the interior cable, which in turn transfers the impact energy through the footings and into their foundation, thereby minimizing the chance that the impact will snap the beam and allow the intruding vehicle to pass through.

Crash beam barriers can employ drop-arm, cantilever, or swing-beam designs. Drop-arm crash beams operate by raising and lowering the beam vertically across the road. Cantilever crash beams are projecting structures that are opened and closed by extending the beam from the hinge buttress to the receiving buttress located on the opposite side of the road. In the swing-beam design, the beam is hinged to the buttress such that it swings horizontally across the road. Generally, swing-beam and cantilever designs are used at locations where a vertical lift beam is impractical. For example, the swing-beam or cantilever designs are utilized at entrances and exits with overhangs, trees, or buildings that would physically block the operation of the drop-arm beam design.

Installing any of these crash beam barriers involves the excavation of a pit approximately 48 inches deep for both the hinge and the receiver footings. Due to the depth of excavation, the site should be inspected for underground utilities before digging begins. Table 9.2 lists the pros and cons of crash beams.

In contrast to wedge barriers and crash beams, which are typically installed separately from a fence line, **gates** are often integrated units of a perimeter fence or wall around a facility.

Gates are basically movable pieces of fencing that can be opened and closed across a road. When the gate is in the closed (armed) position, the leaves of

Table 9.2. Pros and Cons of Crash Beams

Pros	*Cons*
Requires little maintenance, while providing long-term durability.	Crash beams have a slower response time (normally 9.5–15.3 seconds, but can be reduced to 7–10 seconds in emergency situations) than do other types of active security barriers, such as wedge barriers. Because emergency activation of the barrier causes more wear and tear on the system than does normal activation, it is recommended for use only in true emergency situations.
No excavation is required in the roadway itself to install crash beams.	All three crash beam designs possess a low crash rating relative to other types of barriers, such as wedge barriers, and thus they typically are used for lower security applications.
	Certain crash barriers may not be visible to oncoming traffic and therefore may require additional lighting and/or other warning markings to reduce the potential for traffic to accidentally run into the beam.

USEPA (2005).

the gate lock into steel buttresses that are embedded in concrete foundation located on both sides of the roadway, thereby blocking access to the roadway. Generally, gate barricades are constructed from a combination of heavy gauge steel and aluminum that can absorb an impact from vehicles attempting to ram through them. Any remaining impact energy not absorbed by the gate material is transferred to the steel buttresses and their concrete foundation.

Gates can utilize a cantilever, linear, or swing design. Cantilever gates are projecting structures that operate by extending the gate from the hinge footing across the roadway to the receiver footing. A linear gate is designed to slide across the road on tracks via a rack and pinion drive mechanism. Swing gates are hinged so that they can swing horizontally across the road.

Installation of the cantilever, linear, or swing gate designs described above involve the excavation of a pit approximately 48 inches deep for both the hinge and receiver footings to which the gates are attached. Due to the depth of excavation, the site should be inspected for underground utilities before digging begins. Table 9.3 lists the pros and cons of gates.

Table 9.3. Pros and Cons of Gates

Pros	*Cons*
All three gate designs possesses an intermediate crash rating, thereby allowing them to be utilized for medium to higher security applications.	Gates have a slower response time (normally 10–15 seconds, but can be reduced to 7–10 seconds in emergency situations) than do other types of active security barriers, such as wedge barriers. Because emergency activation of the barrier causes more wear and tear on the system than does normal activation, it is recommended for use only in true emergency situations.
Requires very little maintenance.	
Can be tailored to blend in with perimeter fencing.	
Gate construction requires no roadway excavation.	
Cantilever gates are useful for roads with high crowns or drainage gutters.	
These types of barriers are extremely visible, which may deter intruders.	
Gates can also be used to control pedestrian traffic.	

USEPA (2005).

Bollards are vertical barriers at least 3 feet tall and 1 to 2 feet in diameter that are typically set 4 to 5 feet apart from each other so that they block vehicles from passing between them. Smaller bollards, usually four-inch-diameter pipe filled with concrete, are installed in parking areas to prevent vehicles from striking walls or windows or to protect walkway areas. Bollards can either be fixed in place, removable, or retractable. Fixed and removable bollards are passive barriers that are typically used along building perimeters or on sidewalks to block out vehicles, while allowing pedestrians to pass through them. In contrast to passive bollards, retractable bollards are active security barriers that can easily be raised and lowered to allow vehicles to pass between them. Thus, they can be used in driveways or on roads to control vehicular access. When the bollards are raised, they protect above the road surface and block the roadway; when they are lowered, they sit flush with the road surface, and thus allow traffic to pass over them. Retractable bollards are typically constructed from steel or other materials that have a low weight-to-volume ratio so that they require low power to raise and lower. Steel is also more resistant to breaking than is a more brittle material, such as concrete, and is better able to withstand direct vehicular impact without breaking apart.

Retractable bollards are installed in a trench dug across a roadway—typically at an entrance or gate. Installing retractable bollards requires preparing the road surface. Depending on the vendor, bollards can be installed either in a continuous slab of concrete, or in individual excavations with concrete poured in place. The required excavation for a bollard is typically slightly wider and slightly deeper than the bollard height when extended aboveground. The bottom of the excavation is typically lined with gravel to allow drainage. The bollards are then connected to a control panel which controls the raising and lowering of the bollards. Installation typically requires mechanical, electrical, and concrete work; if utility personnel with these skills are available, then the utility can install the bollards themselves. Table 9.4 lists the pros and cons of retractable bollards.

Table 9.4. Pros and Cons of Retractable Bollards

Pros	*Cons*
Bollards have a quick response time (normally 3 to 10 seconds, but can be reduced to 1–3 seconds in emergency situations).	Bollard installations will require construction below the surface of the roadway, which may interfere with buried utilities.
Bollards have an intermediate crash rating, which allows them to be utilized for medium to higher security applications.	Some maintenance is needed to ensure barrier is free to move up and down.
	The distance between bollards must be decreased (i.e., more bollards must be installed along the same perimeter) to make these systems effective against small vehicles (i.e., motorcycles).

USEPA (2005).

Portable/removable barriers, which can include removable crash beams and wedge barriers, are mobile obstacles that can be moved in and out of position on a roadway. For example, a crash beam may be completely removed and stored off-site when it is not needed. An additional example would be wedge barriers that are equipped with wheels that can be removed after the barricade is towed into place.

When portable barricades are needed, they can be moved into position rapidly. To provide them with added strength and stability, they are typically anchored to buttress boxes that are located on either side of the road. These buttress boxes, which may or may not be permanent, are usually filled with sand, water, cement, gravel, or concrete to make them heavy and aid in stabilizing the portable barrier. In addition, these buttresses can help dissipate any impact energy from vehicles crashing into the barrier itself.

Because these barriers are not anchored into the roadway, they do not require excavation or other related construction for installation. In contrast, they can be assembled and made operational in a short period of time. The primary shortcoming to this type of design is that these barriers may move if they are hit by vehicles. Therefore, it is important to carefully assess the placement and anchoring of these types of barriers to ensure that they can withstand the types of impacts that may be anticipated at that location. Table 9.5 lists the pros and cons of portable/removable barricades.

Table 9.5. Pros and Cons of Portable/Removable Barricades

Pros	*Cons*
Installing portable barricades requires no foundation or roadway excavation.	Portable barriers may move slightly when hit by a vehicle, resulting in a lower crash resistance.
Can be moved in and out of position in a short period of time.	Portable barricades typically require 7.75 to 16.25 seconds to move into place, and thus they are considered to have a medium response time when compared with other active barriers.
Wedge barriers equipped with wheels can be easily towed into place.	
Minimal maintenance is needed to keep barriers fully operational.	

USEPA (2005).

Because the primary threat to active security barriers is that vehicles will attempt to crash through them, their most important attributes are their size, strength, and crash resistance. Other important features for an active security barrier are the mechanisms by which the barrier is raised and lowered to allow authorized vehicle entry, and other factors, such as weather resistance and safety features.

Alarms

An *alarm system* is a type of electronic monitoring system that is used to detect and respond to specific types of events—such as unauthorized access to an asset or a possible fire. Note that this discussion does not focus on alarm systems that are not related to a facility's processes. Alarm systems can be integrated with fire detection systems, intrusion-detection systems (IDSs), access control systems, or closed-circuit television (CCTV) systems, such that these systems automatically respond when the alarm is triggered. For example, a smoke detector alarm can be set up to automatically notify the fire department when smoke is detected; or an intrusion alarm can automatically trigger cameras to turn on in a remote location so that personnel can monitor that location.

An alarm system consists of sensors that detect different types of events; an arming station that is used to turn the system on and off; a control panel that receives information, processes it, and transmits the alarm; and an annunciator that generates a visual and/or audible response to the alarm. When a sensor is tripped it sends a signal to a control panel, which triggers a visual or audible alarm and/or notifies a central monitoring station. A more complete description of each of the components of an alarm system is provided below.

Detection devices (also called **sensors**), are designed to detect a specific type of event (such as smoke, intrusion). Depending on the type of event they are designed to detect, sensors can be located inside or outside of the facility or other asset. When an event is detected, the sensors use some type of communication method (such as wireless radio transmitters, conductors, or cables) to send signals to the control panel to generate the alarm. For example, a smoke detector sends a signal to a control panel when it detects smoke.

Alarms use either normally closed (NC) or normally open (NO) electric loops, or "circuits," to generate alarm signals. These two types of circuits are discussed separately below.

In NC loops or circuits, all of the system's sensors and switches are connected in series. The contacts are "at rest" in the closed (on) position, and current continually passes through the system. However, when an event triggers the sensor, the loop is opened, breaking the flow of current through the system and triggering the alarm. NC switches are used more often than are NO switches because the alarm will be activated if the loop or circuit is broken or cut, thereby reducing the potential for circumventing the alarm. This is known as a "supervised" system.

In NO loops or circuits, all of the system's sensors and switches are connected in parallel. The contacts are "at rest" in the open (off) position, and no current passes through the system. However, when an event triggers the sensor, the loop is closed. This allows current to flow through the loop, powering the alarm. NO systems are not "supervised" because the alarm will not be activated if the loop or circuit is broken or cut. However, adding an end-of-line resistor to an NO loop will cause the system to alarm if tampering is detected.

An **arming station**, which is the main user interface with the security system, allows the user to arm (turn on), disarm (turn off), and communicate with the system. How a specific system is armed will depend on how it is used. For example, while IDSs can be armed for continuous operation (twenty-four hours/day), they are usually armed and disarmed according to the work schedule at a specific location so that personnel going about their daily activities do not set off the alarms. In contrast, fire protection systems are typically armed twenty-four hours/day.

A **control panel**, which receives information from the sensors and sends it to an appropriate location, such as to a central operations station or to a 24-hour monitoring facility. Once the alarm signal is received at the central monitoring location, personnel monitoring for alarms can respond (such as by sending security teams to investigate or by dispatching the fire department).

An **annunciator**, which responds to the detection of an event by emitting a signal. This signal may be visual, audible, electronic, or a combination of these three. For example, fire alarm signals will always be connected to audible annunciators, whereas intrusion alarms may not be.

Alarms can be reported locally, remotely, or both locally and remotely. Local and remotely (centrally) reported alarms are discussed in more detail below.

A **local alarm** emits a signal at the location of the event (typically using a bell or siren). A "local only" alarm emits a signal at the location of the event but does not transmit the alarm signal to any other location (i.e., it does not transmit the alarm to a central monitoring location). Typically, the purpose of a "local only" alarm is to frighten away intruders, and possibly to attract the attention of someone who might notify the proper authorities. Because no signal is sent to a central monitoring location, personnel can only respond to a local alarm if they are in the area and can hear and/or see the alarm signal.

Fire alarm systems must have local alarms, including both audible and visual signals. Most fire alarm signal and response requirements are codified in the National Fire Alarm Code, National Fire Protection Association (NFPA) 72. NFPA 72 discusses the application, installation, performance, and maintenance of protective signaling systems and their components. In contrast to fire alarms, which require a local signal when fire is detected, many IDSs do not have a local alert device, because monitoring personnel do not wish to inform potential intruders that they have been detected. Instead, these types of systems silently alert monitoring personnel that an intrusion has been detected, thus allowing monitoring personnel to respond.

In contrast to systems that are set up to transmit "local only" alarms when the sensors are triggered, systems can also be set up to transmit signals to a **central location**, such as to a control room or guard post at the utility, or to a police or fire station. Most fire/smoke alarms are set up to signal both at the location of the event and at a fire station or central monitoring station. Many insurance companies require that facilities install certified systems that include alarm communication to a central station. For example, systems certified by the Underwriters Laboratory (UL) require that the alarm be reported to a central monitoring station.

The main differences between alarm systems lie in the types of event detection devices used in different systems. **Intrusion sensors**, for example, consist of two main categories: perimeter sensors and interior (space) sensors. **Perimeter intrusion sensors** are typically applied on fences, doors, walls, windows, and so on, and are designed to detect an intruder before he/she accesses a protected asset. In contrast, **interior intrusion sensors** are designed to detect an intruder who has already accessed the protected asset (i.e., within a protected room or building). These two types of detection devices can be complementary, and they are often used together to enhance security. For example, a typical intrusion alarm system might employ a perimeter glass-break detector that protects against intruders accessing a room through a window, as well as an ultrasonic interior sensor that detects intruders that have gotten into the room without using the window. Table 9.6 lists and describes types of perimeter and interior sensors.

Table 9.6. Perimeter and Interior Sensors

Type of Perimeter Sensor	*Description*
Foil	Foil is a thin, fragile, lead-based metallic tape that is applied to glass windows and doors. The tape is applied to the window or door, and electric wiring connects this tape to a control panel. The tape functions as a conductor and completes the electric circuit with the control panel. When an intruder breaks the door or window, the fragile foil breaks, opening the circuit and triggering an alarm condition.
Magnetic switches (reed switches)	The most widely used perimeter sensor. They are typically used to protect doors, as well as windows that can be opened (windows that cannot be opened are more typically protected by foil alarms).
Glass break detectors	Placed on glass and sense vibrations in the glass when it is disturbed. The two most common types of glass-break detectors are shock sensors and audio discriminators.

Type of Interior Sensor	*Description*
Passive infrared (PIR)	Presently the most popular and cost effective interior sensors. PIR detectors monitor infrared radiation (energy in the form of heat) and detect rapid changes in temperature within a protected area. Because infrared radiation is emitted by all living things, these types of sensors can be very effective.
Quad PIRs	Consist of two dual-element sensors combined in one housing. Each sensor has a separate lens and a separate processing circuitry, which allows each lens to be set up to generate a different protection pattern
Ultrasonic detectors	Emit high-frequency sound waves, and sense movement in a protected area by sensing changes in these waves. The sensor emits sound waves that stabilize and set a baseline condition in the area to be protected. Any subsequent movement within the protected area by a would-be intruder will cause a change in these waves, thus creating an alarm condition.
Microwave detectors	Emit ultra- high-frequency radio waves, and the detector senses any changes in these waves as they are reflected throughout the protected space. Microwaves can penetrate through walls, and thus a unit placed in one location may be able to protect multiple rooms.
Dual technology devices	Incorporate two different types of sensor technology (such as PIR and microwave technology) together in one housing. When both technologies sense an intrusion, an alarm is triggered.

USEPA (2005).

Fire detection/fire alarm systems consist of different types of fire detection devices and fire alarm systems available. These systems may detect fire, heat, smoke, or a combination of any of these. For example, a typical fire alarm system might consist of heat sensors, which are located throughout a facility and which detect high temperatures or a certain change in temperature over a fixed time period. A different system might be outfitted with both smoke and heat-detection devices. A summary of several different types of fire/smoke/heat-detection sensors is provided in table 9.7.

Once a sensor in an alarm system detects an event, it must communicate an alarm signal. The two basic types of alarm communication systems are hardwired and wireless. Hardwired systems rely on wire that is run from the control panel to each of the detection devices and annunciators. Wireless systems transmit signals from a transmitter to a receiver through the air—primarily using radio or other waves. Hardwired systems are usually lower-cost, more reliable (they are not affected by terrain or environmental factors), and significantly easier to troubleshoot than are wireless systems. However, a major disadvantage of hardwired systems is that it may not be possible to hardwire all locations (for example, it may be difficult to hardwire remote locations). In addition, running wires to their required locations can be both time consuming and costly. The major advantage to using wireless systems is that they can often be installed in areas where hardwired systems are not feasible. However, wireless components can be much more expensive when compared to hardwired systems. In addition, in the past, it has been difficult to perform self-diagnostics on wireless systems to confirm that they are communicating properly with the controller. Presently, the majority of wireless systems incorporate supervising circuitry, which allows the subscriber to know immediately if there is a problem with the system (such as a broken detection device or a low battery), or if a protected door or window has been left open.

Backflow Prevention Devices

Backflow prevention devices are designed to prevent backflow, which is the reversal of the normal and intended direction of water flow in a water system. Backflow is a potential problem in a petrochemical processing system because if incorrectly cross-connected to potable water it can spread contaminated water back through a distribution system. For example, backflow at uncontrolled cross-connections (cross-connections are any actual or potential connection

Table 9.7. Fire/Smoke/Heat Detection Sensors

Detector Type	*Description*
Thermal detector	Sense when temperatures exceed a set threshold (fixed temperature detectors) or when the rate of change of temperature increases over a fixed time period (rate-of-rise detectors).
Duct detector	Is located within the haring and ventilation ducts of the facility. This sensor detects the presence of smoke within the system's return or supply ducts. A sampling tube can be added to the detector to help span the width of the duct.
Smoke detectors	Sense invisible and/or visible products of combustion. The two principle types of smoke detectors are photoelectric and ionization detectors. The major differences between these devices are described below: • Photoelectric smoke detectors react to visible particles of smoke. These detectors are more sensitive to the cooler smoke with large smoke particles that is typical of smoldering fires. • Ionization smoke detectors are sensitive to the presence of ions produced by the chemical reactions that take place with few smoke particle, such as those typically produced by fast burning/flaming fires.
Multi-sensor detectors	Are a combination of photoelectric and thermal detectors. The photoelectric sensor serves to detect smoldering fires, while the thermal detector senses the heat given off from fast burning/flaming fires.
Carbon monoxide (CO) detectors	Are used to indicate the outbreak of fire by sensing the level of carbon monoxide in the air. The detector has an electrochemical cell which senses carbon monoxide, but not some or other products of combustion.
Beam detectors	Are designed to protect large, open spaces such as industrial warehouses. These detectors consist of three parts: the transmitter, which projects a beam of infrared light; the receiver, which registers the light and produces an electrical signal; and the interface, which processes the signal and generates alarm of fault signals. In the event of a fire, smoke particles obstruct the beam of light. Once a preset threshold is exceeded, the detector will go into alarm.
Flame detectors	Sense either ultraviolet (UV) or infrared (IR) radiation emitted by a fire.
Air-sampling detectors	Actively and continuously sample the air from a protected space and are able to sense the pre-combustion stages of incipient fire.

USEPA (2005).

between the public water supply and a source of chemical contamination) or pollution can allow pollutants or contaminants to enter the potable water system. More specifically, backflow from private plumbing systems, industrial areas, hospitals, and other hazardous contaminant-containing systems into public water mains and wells poses serious public health risks and security problems. Cross-contamination from private plumbing systems can contain biological hazards (such as bacteria or viruses) or toxic substances that can contaminate and sicken an entire population in the event of backflow. The majority of historical incidences of backflow have been accidental, but growing concern that contaminants could be intentionally backfed into a system is prompting increased awareness for private homes, businesses, industries, and areas most vulnerable to intentional strikes. Therefore, backflow prevention is a major tool for the protection of water systems.

Backflow may occur under two types of conditions: backpressure and backsiphonage. **Backpressure** is the reverse from normal flow direction within a piping system that is the result of the downstream pressure being higher than the supply pressure. These reductions in the supply pressure occur whenever the amount of water being used exceeds the amount of water supplied, such as during water line flushing, firefighting, or breaks in water mains. **Backsiphonage** is the reverse from normal flow direction within a piping system that is caused by negative pressure in the supply piping (i.e., the reversal of normal flow in a system caused by a vacuum or partial vacuum within the water supply piping). Backsiphonage can occur where there is a high velocity in a pipe line; when there is a line repair or break that is lower than a service point; or when there is lowered main pressure due to high water withdrawal rate, such as during firefighting or water main flushing.

To prevent backflow, various types of backflow preventers are appropriate for use. The primary types of backflow preventers are:

- Air gap drains
- Double-check valves
- Reduced pressure principle assemblies
- Pressure vacuum breakers

Biometric Security Systems

Biometrics involves measuring the unique physical characteristics or traits of the human body. Any aspect of the body that is measurably different from person to person—for example fingerprints or eye characteristics—can serve as a unique biometric identifier for that individual. Biometric systems recognizing fingerprints, palm shape, eyes, face, voice, and signature comprise the bulk of the current biometric systems that recognize other biological features do exist.

Biometric security systems use biometric technology combined with some type of locking mechanisms to control access to specific assets. In order to access an asset controlled by a biometric security system, an individual's biometric trait must be matched with an existing profile stored in a database. If there is a match between the two, the locking mechanisms (which could be a physical lock, such as at a doorway, an electronic lock, such as at a computer terminal, or some other type of lock) is disengaged, and the individual is given access to the asset.

A biometric security system is typically comprised of the following components:

- A sensor, which measures/records a biometric characteristic or trait
- A control panel, which serves as the connection point between various system components. The control panel communicates information back and forth between the sensor and the host computer, and controls access to the asset by engaging or disengaging the system lock based on internal logic and information from the host computer
- A host computer, which processes and stores the biometric trait in a database
- Specialized software, which compares an individual image taken by the sensor with a stored profile or profiles
- A locking mechanism which is controlled by the biometric system
- A power source to power the system.

Biometric Hand and Finger Geometry Recognition

Hand and finger geometry recognition is the process of identifying an individual through the unique "geometry" (shape, thickness, length, width, etc.) of that individual's hand or fingers. Hand geometry recognition has been employed since the early 1980s and is among the most widely used biometric technologies for controlling access to important assets. It is easy to install and use, and is appropriate for use in any location requiring use of two-finger, highly accurate, non-intrusion biometric security. For example, it is currently used in numerous workplaces, daycare facilities, hospitals, universities, airports, refineries, and power plants.

A newer option within hand geometry recognition technology is finger geometry recognition (not to be confused with fingerprint recognition). Finger geometry recognition relies on the same scanning methods and technologies as does hand geometry recognition, but the scanner only scans two of the user's fingers, as opposed to his or her entire hand. Finger geometry recognition has been in commercial use since the mid-1990s and is mainly used in time and attendance applications (i.e., to track when individuals have entered and exited a location). To date the only large-scale commercial use of two finger geometry for controlling access is at Disney World, where season pass holders use the geometry of their index and middle finger to gain access to the facilities.

To use a hand or finger geometry unit, an individual presents his or her hand or fingers to the biometric unit for "scanning." The scanner consists of a Charged Coupled Device (CCD), which is essentially a high-resolution digital camera; a reflective platen on which the hand is placed; and a mirror or mirrors that help capture different angles of the hand or fingers. The camera "scans" individual geometric characteristics of the hand or fingers by taking multiple images while the user's hand rests on the reflective platen. The camera also captures "depth," or three-dimensional information, through light reflected from the mirrors and the reflective platen. This live image is then compared to a "template" that was previously established for that individual when they were "enrolled" in the system. If the live scan of the individual matches the stored template, the individual is "verified," and is given access to that asset. Typically, verification takes about two seconds. In access control applications, the scanner is usually connected to some sort of electronic lock, which unlocks the door, turnstile, or other entry barrier when the user is verified. The user can then proceed through the entrance. In time and attendance applications, the time that an individual checks in and out of a location is stored for later use.

As discussed above, hand and finger geometry recognition systems can be used in several different types of applications, including access control and time and attendance tracking. While time and attendance tracking can be used for security, it is primarily used for operations and payroll purposes (i.e., clocking in and clocking out). In contrast, access control applications are more likely to be security-related. Biometric systems are widely used for access control, and can be used on various types of assets, including entryways, computers, vehicles, and so on. However, because of their size, hand/finger recognition systems are primarily used in entryway access control applications.

Biometric Overview-Iris Recognition

The iris, which is the colored or pigmented area of the eye surrounded by the sclera (the white portion of the eye), is a muscular membrane that controls the amount of light entering the eye by contracting or expanding the pupil (the dark center of the eye). The dense, unique patterns of connective tissue in the human iris were first noted in 1936, but it was not until 1994, when algorithms for iris recognition were created and patented, that commercial applications using biometric iris recognition began to be used extensively. There are now two vendors producing iris-recognition technology: the original developer of these algorithms and a second company that has developed and patented a different set of algorithms to use.

The iris is an ideal characteristic for identifying individuals because it is formed *in utero*, and its unique patterns stabilize around eight months after birth. No two irises are alike; neither an individual's right or left irises, nor the irises of identical twins. The iris is protected by the cornea (the clear covering over the eye), and therefore it is not subject to the aging or physical changes (and potential variation) that are common to some other biometric measures, such as hand, fingerprints, and face. Although some limited changes can occur naturally over time, these changes generally occur in the iris's melanin and therefore affect only the eye's color, and not its unique patterns (in addition, because iris scanning uses only black-and-white images, color changes would not affect the scan anyway). Thus, barring specific injuries or certain rare surgeries directly affecting the iris, the iris's unique patterns remain relatively unchanged over an individual's lifetime.

Iris recognition systems employ a monochromatic or black-and-white video camera that uses both visible and near infrared light to take video of an individual's iris. Video is used rather than still photography as an extra security procedure. The video is used to confirm the normal continuous fluctuations of the pupil as the eye focuses, which ensures that the scan is of a living human being, and not a photograph or some other attempted hoax. A high-resolution image of the iris is then captured or extracted from the video, using a device often referred to as a "frame grabber." The unique characteristics identified in this image are then converted into a numeric code, which is stored as a "template" for that user.

Card Identification/Access/Tracking Systems

A card reader system is a type of electronic identification system that is used to identify a card and then perform an action associated with that card. Depending on the system, the card may identify where a person is or where they were at a certain time; or it may authorize another action, such as disengaging a lock. For example, a security guard may use her card at card readers located throughout a facility to indicate that she has checked a certain location at a certain time. The reader will store this information and/or send it to a central location, where it can be checked later to ensure that the guard has patrolled the area. Other card reader systems can be associated with a lock, so that the cardholder must have their card read and accepted by the reader before the lock disengages.

A complete card reader system typically consists of the following components:

- Access cards that are carried by the user
- Card readers, which read the card signals and send the information to control units
- Control units, which control the response of the card reader to the card
- A power source

A "card" may be a typical card or another type of device, such as a key fob or wand. These cards store electronic information, which can range from a simple code (i.e., the alphanumeric code on a proximity card) to individualized personal data (i.e., biometric data on a smartcard). The card reader reads the information stored on the card and sends it to the control unit, which determines the appropriate action to take when a card is presented. For example, in a card access system, the control unit compares the information on the card vs. stored access authorization information to determine if the card holder is authorized to proceed through the door. If the information stored in the card reader system indicates that the key is authorized to allow entrance through the doorway, the system disengages the lock and the key holder can proceed through the door.

There are many different types of card reader systems on the market. The primary differences between card reader systems are different in the way that data is encoded on the cards and in the way these data are transferred between the card and the card reader, and in the types of applications for which they are best suited. However, all card systems are similar in the way that the card reader and control unit interact to respond to the card.

While card readers are similar in the way that the card reader and control unit interact to control access, they are different in the way data is encoded on the cards and the way these data are transferred between the card and the card reader. There are several types of technologies available for card reader systems. These include:

- Proximity
- Wiegand
- Smartcard
- Magnetic stripe
- Bar code
- Infrared
- Barium ferrite
- Hollerith
- Mixed technologies

Table 9.8 below summarizes various aspects of card reader technologies. The determination for the level of security rate (low, moderate, or high) based on the level of technology a given card reader system has and how simple it is to duplicate that technology, and thus bypass the security. Vulnerability ratings were based on whether the card reader can be damaged easily due to frequent use or difficult working conditions (i.e., weather conditions if the reader is located outside). Often this is influenced by the number of moving parts in the system—the more moving parts, then greater the system's potential susceptibility to damage. The life cycle rating is based on the durability of a given card reader system over its entire operational period. Systems requiring frequent physical contact between the reader and the card often have a shorter life cycle due to the wear and tear to which the equipment is exposed. For many card reader systems, the vulnerability rating and life cycle ratings have a reciprocal relationship. For instance, if a given system has a high vulnerability rating it will almost always have a shorter life cycle.

Table 9.8. Card Reader Technology

Types of Card Readers	*Technology*	*Life Cycle*	*Vulnerability*	*Level of Security*
Proximity	Embedded radio frequency circuits encoded with unique information	Long	Virtually none	Moderate-high
Wiegand	Short lengths of small-diameter, special alloy wire with unique magnetic properties	Long	Low susceptibility to damage; high durability due to embedded wires	Moderate-expensive
Magnetic stripe	Electromagnetic charges to encode information on a piece of tape attached to back of card	Moderate	Moderately susceptible to damage due to frequency of use	Low-Moderate
Bar code	Series of narrow and wide bars and spaces	Short	High; easily damaged	Low
Hollerith	Holes punched in a plastic or paper card and read optically	Short	High; easily damaged from frequent use	Low
Infrared	An encoded shadow pattern within the card, read using an infrared scanner	Moderate	IR scanners are optical and thus, vulnerable to contamination	High
Barium ferrite	Uses small bits of magnetized barium ferrite, placed inside a plastic card. The polarity and location of the "spots" determines the coding	Moderate	Low susceptibility to damage; durable since spots are embedded in the material	Moderate-High
Smartcards	Patterns or series of narrow and wide bars and spaces	Short	High susceptibility to damage, low durability	Highest

USEPA (2005).

Card reader technology can be implemented for facilities of any size and with any number of users. However, because individual systems vary in the complexity of their technology and in the level of security they can provide to a facility, individual users must determine the appropriate system for their needs. Some important features to consider when selecting a card reader system include:

- The technological sophistication and security level of the card system.
- The size and security needs of the facility.
- The frequency with which the card system will be used. For systems that will experience a high frequency of use it is important to consider a system that has a longer life cycle and lower vulnerability rating, thus making it more cost effective to implement.
- The conditions in which the system will be used (i.e., will it be used on the interior or exterior of buildings, does it require light or humidity controls, etc.). Most card reader systems can operate under normal environmental conditions, and therefore this would be a mitigating factor only in extreme conditions.
- System costs.

Exterior Intrusion Sensors

An exterior intrusion sensor is a detection device that is used in an outdoor environment to detect intrusions into a protected area. These devices are designed to detect an intruder, and then communicate an alarm signal to an alarm system. The alarm system can respond to the intrusion in many different ways, such as by triggering an audible or visual alarm signal, or by sending an electronic signal to a central monitoring location that notifies security personnel of the intrusion.

Intrusion sensors can be used to protect many kinds of assets. Intrusion sensors that protect physical space are classified according to whether they protect indoor, or "interior" space (i.e., an entire building or room within a building), or outdoor, or "exterior" space (i.e., a fenceline or perimeter). Interior intrusion sensors are designed to protect the interior space of a facility by detecting an intruder who is attempting to enter, or who has already entered a room or building. In contrast, exterior intrusion sensors are designed to detect an intrusion into a protected outdoor/exterior area. Exterior protected areas are typically arranged as zones or exclusion areas placed so that the intruder is detected early in the intrusion attempt before the intruder can gain access to more valuable assets (e.g., into a building located within the protected area). Early detection creates additional time for security forces to respond to the alarm.

Exterior intrusion sensors are classified according to how the sensor detects the intrusion within the protected area. The three classes of exterior sensor technology include:

- Buried-line sensors
- Fence-associated sensors
- Freestanding sensors

Buried-Line Sensors

As the name suggests, buried line sensors are sensors that are buried underground, and are designed to detect disturbances within the ground—such as disturbances caused by an intruder digging, crawling, walking, or running on the monitored ground. Because they sense ground disturbances, these types of sensors are able to detect sense able to detect intruder activity both on the surface and below ground. Individual types of exterior buried line sensors function in different ways, including, by detecting motion, pressure, or vibrations within the protected ground, or by detecting changes in some type of field (e.g., magnetic field) that the sensors generate within the protected ground. Specific types of buried line sensors include pressure or seismic sensors, magnetic field sensors, ported coaxial cables, and fiber optic cables. Details on each of these sensor types are provided below.

- *Buried-line pressure* or *seismic sensors* detect physical disturbances to the ground—such as vibrations or soil compression—caused by intruders walking, driving, digging, or otherwise physically contacting the protected ground. These sensors detect disturbances from all directions and, therefore, can protect an area radially outward from their location; however, because detection may weaken as a function of distance from the disturbance, choosing the correct burial depth from the design area will be crucial. In general, sensors buried at a shallow depth protect a relatively small area but have a high probability of detecting intrusion within that area, while sensors buried at a deeper depth protect a wider area but have a lower probability of detecting intrusion into that area.
- *Buried-line magnetic field sensors* detect changes in a local magnetic field that are caused by the moment of metallic objects within that field. This type of sensor can detect ferric metal objects worn or carried by an intruder entering a protected area on foot as well as vehicles being driven into the protected area.
- *Buried-line ported coaxial cable sensors* detect the motion of any object (i.e., human body, metal, etc.) possessing high conductivity and located within

close proximity to the cables. An intruder entering into the protected space creates an active disturbance in the electric field, thereby triggering an alarm condition.

- *Buried-line fiber optic cable sensors* detect changes in the attenuation of light signals transmitted within the cable. When the soil around the cable is compressed, the cable is distorted, and the light signal transmitted through the cable changes, initiating an alarm. This type of sensor is easy to install because it can be buried at a shallow burial depth (only a few centimeters) and still be effective.

Table 9.9 presents the distinctions between the four types of buried sensors.

Table 9.9. Types of Buried Sensors

Type	*Description*
Pressure or seismic	Responds to disturbances in the soil
Magnetic field	Responds to a change in the local magnetic field caused by the movement of nearby metallic material
Ported coaxial cables	Responds to motion of a material with a high dielectric constant or high conductivity near the cables
Fiber optic cables	Responds to a change in the shape of the fiber that can be sensed using sophisticated sensors and computer signal processing

Adapted from Garcia, M. L., 2001.

Fence-Associated Sensors

Fence-associated sensors are either attached to an existing fence, or are installed in such a way as to create a fence. These sensors detect disturbances to the fence—such as those caused by an intruder attempting to climb the fence, or by an intruder attempting to cut or lift the fence fabric. Exterior fence-associated sensors include fence-disturbance sensors, taut-wire sensor fences, and electric field or capacitance sensors. Details on each of these sensor types are provided below.

- *Fence-disturbance sensors* detect the motion or vibration of a fence, such as that can be caused by an intruder attempting to climb or cut through the fence. In general, fence disturbance sensors are used on chain link fences or on other fence types where a moveable fence fabric is hung between fence posts.

- *Taut-wire sensor fences* are similar to fence-disturbance sensors except that instead of attaching the sensors to a loose fence fabric, the sensors are attached to a wire that is stretched tightly across the fence. These types of systems are designed to detect changes in the tension of the wire rather than vibrations in the fence fabric. Taut-wire sensor fences can be installed over existing fences, or as stand-alone fence systems.
- *Electric field or capacitance sensors* detect changes in capacitive coupling between wires that are attached to, but electrically isolated from, the fence. As opposed to other fence-associated intrusion sensors, both electric field and capacitance sensors generate an electric field that radiates out from the fence line, resulting in an expanded zone of protection relative to other fence-associated sensors, and allowing the sensor to detect an intruders' presence before they arrive at the fence line. Note: proper spacing is necessary during installation of the electric field sensor to detect a would-be intruder from slipping between largely spaced wires.

Free-Standing Sensors

These sensors, which include active infrared, passive infrared, bistatic microwave, monostatic microwave, dual-technology, and video motion detection (VMD) sensors, consist of individual sensor units or components that can be set up in a variety of configurations to meet a user's needs. They are installed above-ground, and depending on how they are oriented relative to each other, they can be used to establish a protected perimeter or a protected space. More details on each of these sensor types are provided below.

- *Active infrared sensors* transmit infrared energy into the protected space, and monitor for changes in this energy caused by intruders entering that space. In a typical application, an infrared light beam is transmitted from a transmitter unit to a receiver unit. If an intruder crosses the beam, the beam is blocked, and the receiver unit detects a change in the amount of light received, triggering an alarm. Different sensors can see single- and multiple-beam arrays. Single-beam infrared sensors transmit a single infrared beam. In contrast, multiple-beam infrared sensors transmit two or more beams parallel to each other. This multiple-beam sensor arrangement creates an infrared "fence."
- *Passive infrared (PIR) sensors* monitor the ambient infrared energy in a protected area, and evaluate changes in that ambient energy that may be caused

by intruders moving through the protected area. Detection ranges can exceed 100 yards on cold days with size and distance limitations dependent upon the background temperature. PIR sensors generate a non-uniform detection pattern (or "curtain") that has areas (or "zones") of more sensitivity and areas of less sensitivity. The specific shape of the protected area is determined by the detector's lenses. The general shape common to many detection patterns is a series of long "fingers" emanating from the PIR and spreading in various directions. When intruders enter the detection area, the PIR sensor detects differences in temperature due to the intruder's body heat, and triggers an alarm. While the PIR leaves unprotected areas between its fingers, an intruder would be detected if he passed from a non-protected area to a protected area.

- *Microwave sensors* detect changes in received energy generated by the motion of an intruder entering into a protected area. Monostatic microwave sensors incorporate transmitter and a receiver in one unit, while bistatic sensors separate the transmitter and the receiver into different units. Monostatic sensors are limited to a coverage area of 400 feet, while bistatic sensors can cover an area up to 1,500 feet. For bistatic sensors, a zone of no detection exists in the first few feet in front of the antennas. This distance from the antennas to the point at which the intruder is first detected is known as the offset distance. Due to this offset distance, antennas must be configured so that they overlap one another (as opposed to being adjacent to each other), thereby creating long perimeters with a continuous line of detection.
- *Dual-technology sensors* consist of two different sensor technologies incorporated together into one sensor unit. For example, a dual technology sensor could consist of a passive infrared detector and a monostatic microwave sensor integrated into the same sensor unit.
- *Video motion detection* (VMD) *sensors* monitor video images from a protected area for changes in the images. Video cameras are used to detect unauthorized intrusion into the protected area by comparing the most recent image against a previously established one. Cameras can be installed on towers or other tall structures so that they can monitor a large area.

Fences

A fence is a physical barrier that can be set up around the perimeter of an asset. Fences often consist of individual pieces (such as individual pickets in a wooden fence, or individual sections of a wrought-iron fence) that are fastened together. Individual sections of the fence are fastened together using posts, which are sunk into the ground to provide stability and strength for the sections of the fence hung between them. Gates are installed between individual sections of the fence to allow access inside the fenced area.

Many fences are used as decorative architectural features to separate physical spaces for each other. They may also be used to physically mark the location of a boundary (such as a fence installed along a properly line). However, a fence can also serve as an effective means for physically delaying intruders from gaining access to a DIB asset. For example, many utilities install fences around their primary facilities, around remote pump stations, or around hazardous petrochemical materials storage areas or sensitive areas within a facility. Access to the area can be controlled through security at gates or doors through the fence (for example, by posting a guard at the gate or by locking it). In order to gain access to the asset, unauthorized persons could either have to go around or through the fence.

Fences are often compared with walls when determining the appropriate system for perimeter security. While both fences and walls can provide adequate perimeter security, fences are often easier and less expensive to install than walls. However, they do not usually provide the same physical strength that walls do. In addition, many types of fences have gaps between the individual pieces that make up the fence (i.e., the spaces between chain links in a chain link fence or the space between pickets in a picket fence). Thus, many types of fences allow the interior of the fenced area to be seen. This may allow intruders to gather important information about the locations or defenses of vulnerable areas within the facility.

There are numerous types of materials used to construct fences, including chain link iron, aluminum, wood, or wire. Important security attributes of a fence include the height to which it can be constructed, the strength of the material comprising the fence, the method and strength of attaching the individual sections of the fence together at the posts and the fence's ability to restrict the view of the assets inside the fence. Additional considerations should include the ease of installing the fence and the ease of removing and reusing sections of the fence. Table 9.10 provides a comparison of the important security and usability features of various fence types.

Table 9.10. Comparison of Different Fence Types

Specifications	*Chain Link*	*Iron*	*Wire (Wirewall)*	*Wood*
Height limitations	12'	12'	12'	8'
Strength	Medium	High	High	Low
Installation requirements	Low	High	High	Low
Ability to remove/reuse	Low	High	Low	High
Ability to replace/repair	Medium	High	Low	High

USEPA (2005).

Some fences can include additional measures to delay, or even detect, potential intruders. Such measures may include the addition of barbed wire, razor wire, or other deterrents at the top of the fence. Barbed wire is sometimes employed at the base of fences as well. This can impede a would-be intruder's progress in even reaching the fence. Fences may also be fitted with security cameras to provide visual surveillance of the perimeter. Finally, some facilities have installed motion sensors along their fences to detect movement on the fence. Several manufacturers have combined these multiple perimeter security features into one product and offer alarms and other security features.

The correct implementation of a fence can make it a much more effective security measure. Security experts recommend the following when a facility constructs a fence:

- The fence should be at least 7–9 feet high.
- Any outriggers, such as barbed wire, that are affixed on top of the fence should be angled out and away from the facility, and not in toward the facility. This will make climbing the fence more difficult, and will prevent ladders from being placed against the fence.
- Other types of hardware can increase the security of the fence. This can include installing concertina wire along the fence (this can be done in front of the fence or at the top of the fence), or adding intrusion sensors, camera, or other hardware to the fence.
- All undergrowth should be cleared for several feet (typically 6 feet) on both sides of the fence. This will allow for a clearer view of the fence by any patrols in the area.
- Any trees with limbs or branches hanging over the fence should be trimmed so that intruders cannot use them to go over the fence. Also, it should be

noted that fallen trees can damage fences, and so management of trees around the fence can be important. This can be especially important in areas where fence goes through a remote area.

- Fences that do not block the view from outside the fence to inside the fence allow patrols to see inside the fence without having to enter the facility.
- "No Trespassing" signs posted along fence can be a valuable tool in prosecuting any intruders who claim that the fence was broken and that they did not enter through the fence illegally. Adding signs that highlight the local ordinances against trespassing can further persuade simple troublemakers for illegally jumping/climbing the fence. Electrical substation and other electrical component installations should have clearly visible signage warning of HIGH VOLTAGE and the dangers of electrical shock.

Films for Glass-Shatter Protection

Most DIB sector facilities have numerous windows on the outside of buildings, in doors, and in interior offices. In addition, many facilities have glass doors or other glass structures, such as glass walls or display cases. These glass objects are potentially vulnerable to shattering when heavy objects are thrown or launched at them, when explosions occur near them, or when there are high winds (for exterior glass). If the glass is shattered, intruders may potentially enter an area. In addition, shattered glass projected into a room from an explosion or from an object being thrown through a door or window can injure and potentially incapacitate personnel in the room. Materials that prevent glass from shattering can help to maintain the integrity of the door, window, or other glass object, and can delay an intruder from gaining access. These materials can also prevent flying glass and thus reduce potential injuries.

Materials designed to prevent glass from shattering include specialized films and coatings. These materials can be applied to existing glass objects to improve their strength and their ability to resist shattering. The films have been tested against many scenarios that could result in glass breakage, including penetration by blunt objects, bullets, high winds, and simulated explosions. Thus, the films are tested against both simulated weather scenarios (which could include both the high winds themselves and the force of objects blown into the glass), as well as more criminal/terrorist scenarios where the glass is subject to explosives or bullets.

The primary attributes of films for shatter protection are:

- The materials from which the film is made
- The adhesive that bonds the film to the glass surface
- The thickness of the film

Standard glass safety films are designed from high-strength polyester. Polyester provides both strength and elasticity, which is important in absorbing the impact of an object, spreading the force of the impact over the entire film, and resisting tearing. The polyester is also designed to be resistant to scratching, which can result when films are cleaned with abrasives or other industrial cleaners.

The bonding adhesive is important in ensuring that the film does not tear away from the glass surface. This can be especially important when the glass is broken, so that the film does not peel off the glass and allow it to shatter. In addition, films applied to exterior windows can be subject to high concentrations of UV light, which can break down bonding materials.

Film thickness is measured in gauge or mils. According to test results reported by several manufacturers, film thickness appears to affect resistance to penetration/tearing, with thicker films being more resistant to penetration and tearing. However, the appreciation of a thicker film did not decrease glass fragmentation.

Many manufacturers offer films in different thicknesses. The "standard" film is usually one 4 mil layer; thicker films are typically composed of several layers of the standard 4 mil sheet. However, newer technologies have allowed the polyester to be "microlayered" to produce a stronger film without significantly increasing its thickness. In this microlayering process, each laminate film is composed of multiple micro-thin layers of polyester woven together an alternating angles. This provides increased strength for the film, while maintaining the flexibility and thin profile of one film layer.

As described above, many vendors test their products in various scenarios that would lead to glass shattering, including simulated bomb blasts and simulation of the glass being struck by wind-blown debris. Some manufacturers refer to the Government Services Administration standard for bomb blasts, which require resistance to tearing for a 4 PSI blast. Other manufacturers use other measures and test for resistance to tearing. Many of these tests are not "standard," in that no standard testing or reporting methods have been adopted by any of the accepted standards-setting institutions. However, many of the vendors publish the procedure and the results of these tests on their websites, and this may allow users to evaluate the protectiveness of these films. For example, several vendors evaluate the "protectiveness" of their films and the "hazard" resulting from blasts near windows with and without protective films. Protectiveness is usually evaluated based on the percentage of glass ejected from the window, and the height at which that ejected glass travels during the blast (for example, if the blasted glass tends to project upward into a room—potentially toward people's faces—it is a higher hazard than if it is blown downward into the room toward people's feet).

There are some standard measures of glass breakage. For example, several vendors indicated that their products exceed the American Society for Testing and Materials (ASTM) standard 64Z-95 "Standard Test Method for Glazing and Glazing Systems Subject to Air Blast Loadings." Vendors often compare the results of some sort of penetration or force test, ballistic tests, or simulated explosions with unprotected glass versus glass onto which their films have been applied. Results generally show that applying films to the glass surfaces reduces breakage/penetration of the glass and can reduce the amount and direction of glass ejected from the frame. This in turn reduces the hazard from flying glass.

In addition to these types of tests, many vendors conduct standard physical tests on their products, such as tests for tensile strength and peel strength. Tensile strength indicates the strength per area of material, while the peel strength indicates the force it would take to peel the product from the glass surface. Several vendors indicate that their products exceed American National Standards Institute (ANSI) standard Z97.1 for tensile strength and adhesion.

Vendors typically have a warranty against peeling or other forms of deterioration of their products. However, the warranty requires that the films be installed by manufacturer-certified technicians to ensure that they are applied correctly and therefore that the warranty is in effect. Warranties from different manufacturers may vary. Some may cover the cost of replacing the material only, while others include material plus installation. Because installation costs are significantly greater than material costs, different warranties may represent large difference in potential costs.

Fire Hydrant Locks

Fire hydrants are installed at strategic locations throughout a community's water distribution system to supply water for firefighting. However, because there are many hydrants in a system and they are often located in residential neighborhood, industrial districts, and other areas where they cannot be easily observed and/or guarded, they are potentially vulnerable to unauthorized access. Many municipalities, states, and EPA regions have recognized this potential vulnerability and have instituted programs to lock hydrants. For example, EPA Region 1 has included locking hydrants as number 7 on its "Drinking Water Security and Emergency Preparedness" Top-Ten List for small groundwater suppliers.

A **hydrant lock** is a physical security device designed to prevent unauthorized access to the water supply through a hydrant. They can also ensure water and water pressure availability to firefighters and prevent water theft and associated lost water revenue. These locks have been successfully used in numerous municipalities and in various climates and weather conditions.

Fire hydrant locks are basically steel covers or caps that are locked in place over the operating nut of a fire hydrant. The lock prevents unauthorized persons from accessing the operating nut and opening the fire hydrant valve. The lock also makes it more difficult to remove the bolts from the hydrant and access the system that way. Finally, hydrant locks shield the valve from being broken off. Should a vandal attempt to breach the hydrant lock by force and succeed in breaking the hydrant lock, the vandal will only succeed in bending the operating valve. If the hydrant's operating valve is bent, the hydrant will not be operational, but the water asset remains protected and inaccessible to vandals. However, the entire hydrant will need to be replaced.

Hydrant locks are designed so that the hydrants can be operated by special key wrenches without removing the lock. These specialized wrenches are generally distributed to the fire department, public works department, and other authorized persons so that they can access the hydrants as needed. An inventory of wrenches and their serial numbers is generally kept by a municipality so that the location of all wrenches is known. These operating key wrenches may only be purchased by registered lock owners.

The most important features of hydrant are their strength and the security of their locking systems. The locks must be strong so that they cannot be broken off. Hydrant locks are constructed from stainless or alloyed steel.

Stainless-steel locks are stronger and are ideal for all climates; however, they are more expensive than alloy locks. The locking mechanisms for each fire hydrant locking system ensure that the hydrant can only be operated by authorized personnel who have the specialized key to work the hydrant.

Hatch Security

A hatch is basically a door that is installed on a horizontal plane (such as in a floor, a paved lot, or a ceiling), instead of on a vertical plane (such as in a building wall). Hatches are usually used to provide access to assets that are either located underground (such as hatches to basements or underground vaults and storage areas), or to assets located above ceilings (such as emergency roof exits). Securing a hatch by locking it or upgrading materials to give the hatch added strength can help to delay unauthorized access to any asset behind the hatch.

Like all doors, a hatch consists of a frame anchored to the horizontal structure, a door or doors, hinges connecting the door/doors to the frame, and a latching or locking mechanism that keeps the hatch door/doors closed.

It should be noted that improving hatch security is straightforward, and that hatches with upgraded security features can be installed new or they can be retrofit for existing applications.

Depending on the application, the primary security-related attributes of a hatch are the strength of the door and frame, its resistance to the elements and corrosion, its ability to be sealed against water or gas, and its locking features.

Hatches must be both strong and lightweight so that they can withstand typical static loads (such as people or vehicles walking or driving over them) while still being easy to open.

In addition, because hatches are typically installed at outdoor locations, they are usually designed from corrosion-resistant metal that can withstand the elements. Therefore, hatches are typically constructed from high-gauge steel or lightweight aluminum.

Aluminum is typically the material of choice for hatches because it is lightweight and more corrosion resistant relative to steel. Aluminum is not as rigid as steel, so aluminum hatch doors may be reinforced with aluminum stiffeners to provide extra strength and rigidity. The doors are usually constructed from single or double layers (or "leaves") of material. Single-leaf designs are standard for smaller hatches, while double-leaf deigns are required for larger hatches. In addition, aluminum products do not require painting. This reflected in the warranties available with different products. Product warranties range from ten years to lifetime.

Steel is heavier per square foot than aluminum, and thus steel hatches will be heavier and more difficult to open than aluminum hatches of the same size. However, heavy steel hatch doors may have spring-loaded, hydraulic, or gas openers or other specialized features that help in opening the hatch and in keeping it open.

Many hatches are installed in outdoor areas, often in roadways or pedestrian areas. Therefore, the hatch installed for any given application must be designed to withstand the expected load at that location. Hatches are typically solid to withstand either pedestrian of vehicle loading. Pedestrian loading hatches are designed to withstand either 150 or 300 pounds per square feet (psf) of loading. The vehicle loading standard is the American Association of State Highway and Transportation Officials (AASHTO) H-20 wheel loading standard of 16,000 pounds over an 8-inch by 20-inch area. It should be noted that these design parameters are for static loads and not dynamic loads; thus, the loading capabilities may not reflect potential resistance to other types of loads that may be more typical of an intentional threat, such as repeated blows from a sledge hammer or pressure generated by bomb blasts or bullets.

The typical design for a watertight hatch includes a channel frame that directs water away from the hatch. This can be especially important in a hatch on a storage tank because this will prevent liquid contaminants from being dumped on the hatch and leaking through into the interior. Hatches can also be constructed with gasket seals that are air-, odor-, and gas-tight.

Typically, hatches for pedestrian loading applications have hinges located on the exterior of the hatch, while hatches designed for H-20 loads have hinges located in the interior of the hatch. Hinges located on the exterior of the hatch may be able to be removed; thereby allowing intruders to remove the hatch door and access the asset behind the hatch. Therefore, installing H-20 hatches even for applications which do not require H-20 loading levels may increase security, because intruders will not be able to tamper with the hinges and circumvent the hatch this way.

In addition to the location of the hinges, stock hinges can be replaced with heavy duty or security hinges that are more resistant to tampering.

The hatch locking mechanism is perhaps the most important part of hatch security. There are a number of locks that can be implemented for hatches, including:

- Slam locks (internal locks that are located within the hatch frame)
- Recessed cylinder locks
- Bolt locks
- Padlocks

Ladder Access Control

DIB sector facilities have a number of assets that are raised above ground level, including electrical substations, pumping stations, gas compression facilities, raised petrochemical tanks, raised piping systems, and roof access points into buildings. In addition, communications equipment, antennae, or other electronic devices may be located on the top of these raised assets. Typically, these assets are reached by ladders that are permanently anchored to the asset. For example, raised petrochemical/water tanks typically are accessed by ladders that are bolted to one of the legs of the tank. Controlling access to these raised assets by controlling access to the ladder can increase security at a DIB sector facility.

A typical ladder access control system consists of some type of cover that is locked or secured over the ladder. The cover can be a casing that surrounds most of the ladder, or a door or shield that covers only part of the ladder. In either case, several rungs of the ladder (the number of rungs depends on the size of the cover) are made inaccessible by the cover, and these rungs can only be accessed by opening or removing the cover. The cover is locked so that only authorized personnel can open or remove it and use the ladder. Ladder access controls are usually installed at several feet above ground level, and they usually extend several feet up the ladder so that they cannot be circumvented by someone accessing the ladder above the control system.

The important features of ladder access control are the size and strength of the cover and its ability to lock or otherwise be secured from unauthorized access.

The covers are constructed from aluminum or some type of steel. This should provide adequate protection from being pierced or cut through. The metals are corrosion resistant so that they will not corrode or become fragile from extreme weather conditions in outdoor applications. The bolts used to install each of these systems are galvanized steel. In addition, the bolts for each cover are installed on the inside of the unit so they cannot be removed from the outside.

Locks

A lock is a type of physical security device that can be used to delay or prevent a door, a gate, a window, a manhole, a filing cabinet drawer, or some other physical feature from being opened, moved, or operated. Locks typically operate by connecting two pieces together—such as by connecting a door to a door jamb or a manhole to its casement. Every lock has two modes—engaged (or "locked"), and disengaged (or "opened"). When a lock is disengaged, the asset on which the lock is installed can be accessed by anyone, but when the lock is engaged, only approved personnel gain access to the locked asset.

Before discussing locks and their applicability it is important to discuss key control. Based on my experience, many DIB sector facilities (and others) have no idea how many keys for various site/equipment locks have been issued to employees over the years. Many facilities simply issue keys to employees at hiring with no accountability for the keys upon the employee's departure. Needless to say this is not good security policy. You can have the best locks available installed throughout your facilities, but if you do not have proper key control, you do not have proper security.

Locks are excellent security features because they have been designed to function in many ways and to work on many different types of assets. Locks can also provide different levels of security depending on how they are designed and implemented. The security provided by a lock is dependent on several factors, including its ability to withstand physical damage (i.e., can it be cut off, broken, or otherwise physically disabled) as well as its requirements for supervision or operation (i.e., combinations may need to be changed frequently so that they are not compromised and the locks remain secure). While there is no single definition of the "security" of a lock, locks are often described as minimum, medium, or maximum security. Minimum-security locks are those that can be easily disengaged (or "picked") without the correct key or code, or those that can be disabled easily (such as small padlocks that can be cut with bolt cutters). Higher-security locks are more complex and thus are more difficult to pick, or are sturdier and more resistant to physical damage.

Many locks only need to be unlocked from one side. For example, most door locks need a key to be unlocked only from the outside. A person opens such devices, called single-cylinder locks, from the inside by pushing a button or by turning a knob or handle. Double-cylinder locks require a key to be locked or unlocked from both sides.

Manhole Intrusion Sensors

Manholes are commonly found in DIB sector sites. Manholes are designed to provide access to the underground utilities, meter vaults, petrochemical pumping rooms, and so on, and therefore they are potential entry points to a system. Because many utilities run under other infrastructure (roads, buildings), manholes also provide potential access points to critical infrastructure as well as petrochemical process assets. In addition, because the portion of the system to which manholes provide entry is primarily located underground, access to a system through a manhole increases the chance that an intruder will not be seen. Therefore, protecting manholes can be a critical component of guarding an entire site and a surrounding community.

There are multiple methods for protecting manholes, including preventing unauthorized personnel from physically accessing the manhole and detecting attempts at unauthorized access to the manhole. A manhole intrusion sensor is a physical security device designed to detect unauthorized access to the facility through a manhole. Several different technologies have been used to develop manhole intrusion sensors, including mechanical systems, magnetic systems, and fiber optic and infrared sensors. Some of these intrusion sensors have been specifically designed for manholes, while others consist of standard, off-the-shelf intrusion sensors that have been implemented in a system specifically designed for application in a manhole.

Radiation Detection Equipment for Monitoring Personnel and Packages

Radioactive substances brought on-site at a facility could be used to contaminate the facility, thereby preventing personnel from safely entering the facility. In addition, radioactive substances brought on-site at a DIB sector entity could be discharged into the waste system, contaminating the downstream water supply. Therefore, detection of radioactive substances being brought on-site is an important security enhancement.

Different radionuclides have unique properties, and different equipment is required to detect different types of radiation. However, it is impractical and potentially unnecessary to monitor for specific radionuclides being brought on-site. Instead, it may be more useful to monitor for gross radiation as an indicator of unsafe substances.

In order to protect against these radioactive materials being brought on-site, a facility may set up monitoring sites outfitted with radiation detection instrumentation at entrances to the facility. Depending on the specific types of equipment chosen, this equipment would detect radiation emitted from people, packages, or other objects being brought through an entrance.

One of the primary differences between the different types of detection equipment is the means by which the equipment reads the radiation. Radiation may either be detected by direct measurement or through sampling.

Direct radiation measurement involves measuring radiation through an external probe on the detection instrumentation. Some direct measurement equipment detects radiation emitted into the air around the monitored object. Because this equipment detects radiation in the air, it does not require that the monitoring equipment make physical contact with the monitored object. Direct means for detecting radiation include using a walk-through portal-type monitor that would detect elevated radiation levels on a person or in a package, or by using a hand-held detector, which would be moved or swept over individual objects to locate a radioactive source.

Some types of radiation, such as alpha or low-energy beta radiation, have a short range and are easily shielded by various materials. These types of radiation cannot be measured through direct measurement—they must be measured through sampling. Sampling involves wiping the surface to be tested with a special filter cloth, and then reading the cloth in a special counter. For example, specialized smear counters measure alpha and low-energy beta radiation.

Reservoir Covers

Reservoirs are used to store raw or untreated water for petrochemical processing. They can be located underground (buried), at ground level, or on an elevated surface. Reservoirs can vary significantly in size; small reservoirs can hold as little as 1,000 gallons, while larger reservoirs may hold many millions gallons.

Reservoirs can be either natural or man-made. Natural reservoirs can include lakes or other contained water bodies, while man-made reservoirs usually consist of some sort of engineered structure, such as a tank or other impoundment structure. In addition to the water containment structure itself, reservoir systems may also include associated water treatment and distribution equipment, including intakes, pumps, pump houses, piping systems, chemical treatment and chemical storage areas, and so forth.

One of the most serious potential threats to the system is direct contamination of the stored water through dumping contaminants into the reservoir. DIB sector facilities have taken various measures to mitigate this type of threat, including fencing off the reservoir, installing cameras to monitor for intruders, and monitoring for changes in water quality. Another option for enhancing security is covering the reservoir using some type of manufactured cover to prevent intruders from gaining physical access to the stored water. Implementing a reservoir cover may or may not be practical depending on the size of the reservoir (for example, covers are not typically used on natural reservoirs because they are too large for the cover to be technically feasible and cost effective).

A reservoir cover is a structure installed on or over the surface of the reservoir to minimize water quality degradation. The three basic design types for reservoir covers are:

- Floating
- Fixed
- Air-supported

A variety of materials are used when manufacturing a cover, including reinforced concrete, steel, aluminum, polypropylene, chlorosulfonated polyethylene, or ethylene interpolymer alloys. There are several factors that affect a reservoir cover's effectiveness, and thus its ability to protect the stored water. These factors include:

- The location, size, and shape of the reservoir
- The ability to lay/support a foundation (for example, footing, soil, and geotechnical support conditions)

- The length of time reservoir can be removed from service for cover installation or maintenance
- Aesthetic considerations
- Economic factors, such as capital and maintenance costs

For example, it may not be practical to install a fixed cover over a reservoir if the reservoir is too large or if the local soil conditions cannot support a foundation. A floating or air-supported cover may be more appropriate for these types of applications.

In addition to the practical considerations for installation of these types of covers, there are a number of operations and maintenance (O&M) concerns that affect the utility of a cover for specific applications, including how different cover materials will withstand local climatic conditions, what types of cleaning and maintenance will be required for each particular type of cover, and how these factors will affect the covers lifespan and its ability to be repaired when it is damage.

The primary feature affecting the security of a reservoir cover is its ability to maintain its integrity. Any type of cover, no matter what its construction material, will provide good protection from contamination by rainwater or atmospheric deposition, as well as from intruders attempting to access the stored water with the intent of causing intentional contamination. The covers are large and heavy, and it is difficult to circumvent them to get into the reservoir. At the very least, it would take a determined intruder, as opposed to a vandal, to defeat the cover.

Security for Doorways—Side Hinged Doors

Doorways are the main access points to a facility or to rooms within a building. They are used on the exterior or in the interior of buildings to provide privacy and security for the areas behind them. Different types of doorway security systems may be installed in different doorways depending on the needs or requirements of the buildings or rooms. For example, exterior doorways tend to have heavier doors to withstand the elements and to provide some security to the entrance of the building. Interior doorways in office areas may have lighter doors that may be primarily designed to provide privacy rather than security. Therefore, these doors may be made of glass or lightweight wood. Doorways in industrial areas may have sturdier doors than do other interior doorways and may be designed to provide protection or security for areas behind the doorway. For example, fireproof doors may be installed in chemical storage areas or in other areas where there is a danger of fire.

Because they are the main entries into a facility or a room, doorways are often prime targets for unauthorized entry into a facility or an asset. Therefore, securing doorways may be a major step in providing security at a facility.

A doorway includes four main components:

- The door, which blocks the entrance. The primary threat to the actual door is breaking or piercing through the door. Therefore, the primary security features of doors are their strength and resistance to various physical threats, such as fire or explosions.
- The door frame, which connects the door to the wall. The primary threat to a door frame is that the door can be pried away from the frame. Therefore, the primary security feature of a door frame is its resistance to prying.
- The hinges, which connect the door to the door frame. The primary threat to door hinges is that they can be removed or broken, which will allow intruders to remove the entire door. Therefore, security hinges are designed to be resistant to breaking. They may also be designed to minimize the threat of removal from the door.
- The lock, which connects the door to the door frame. Use of the lock is controlled through various security features, such as keys, combinations, etc., such that only authorized personnel can open the lock and go through the door. Locks may also incorporate other security features, such as software or other systems to track overall use of the door or to track individuals using the door, and so on.

Each of these components is integral in providing security for a doorway, and upgrading the security of only one of these components while leaving the other components unprotected may not increase the overall security of the doorway. For example, many facilities upgrade door locks as a basic step in increasing the security of a facility. However, if the facilities do not also focus on increasing security for the door hinges or the door frame, the door may remain vulnerable to being removed from its frame, thereby defeating the increased security of the door lock.

The primary attribute for the security of a door is its strength. Many security doors are 4–20 gauge hollow metal doors consisting of steel plates over a hollow cavity reinforced with steel stiffeners to give the door extra stiffness and rigidity. This increases resistance to blunt force used to try to penetrate through the door. The space between the stiffeners may be filled with specialized materials to provide fire, blast, or bullet resistance to the door.

The Windows and Doors Manufacturers Association has developed a series of performance attributes for doors. These include:

- Structural resistance
- Forced entry resistance
- Hinge style screw resistance
- Split resistance
- Hinge resistance
- Security rating
- Fire resistance
- Bullet resistance
- Blast resistance

The first five bullets provide information on a door's resistance to standard physical breaking and prying attacks. These tests are used to evaluate the strength of the door and the resistance of the hinges and the frame in a standardized way. For example, the Rack Load Test simulates a prying attack on a corner of the door. A test panel is restrained at one end, and a third corner is supported. Loads are applied and measured at the fourth corner. The Door Impact Test simulates a battering attack on a door and frame using impacts of 200-foot pounds by a steel pendulum. The door must remain fully operable after the test. It should be noted that door glazing is also rated for resistance to shattering, and so on. Manufacturers will be able to provide security ratings for these features of a door as well.

Door frames are an integral part of doorway security because they anchor the door to the wall. Door frames are typically constructed from wood or steel, and they are installed such that they extend for several inches over the doorway that has been cut into the wall. For added security, frames can be designed to have varying degrees of overlap with, or wrapping over, the underlying wall. This can make prying the frame from the wall more difficult.

A frame formed from a continuous piece of metal (as opposed to a frame constructed from individual metal pieces) will prevent prying between pieces of the frame.

Many security doors can be retrofit into existing frames; however, many security door installations including replacing the door frame as well as the door itself. For example, bullet-resistance per Underwriter's Laboratory (UL) 752 requires resistance of the door and frame assembly, and thus replacing the door only would not meet UL 752 requirements.

Valve Lockout Devices

Valves are utilized as control elements in fuel oil/natural gas and petrochemical process piping networks. They regulate the flow of both liquids and gases by opening, closing, or obstructing a flow passageway. Valves are typically located where flow control is necessary. They can be located in-line or at pipeline and tank entrance and exit points. They can serve multiple purposes in a process pipe network, including:

- Redirecting and throttling flow
- Preventing backflow
- Shutting off flow to a pipeline or tank (for isolation purposes)
- Releasing pressure
- Draining extraneous liquid from pipelines or tanks
- Introducing chemicals into the process network
- Serving as access points for sampling process water

Valves may be located either aboveground or below ground. It is critical to provide protection against valve tampering. For example, tampering with a pressure relief valve could result in a pressure buildup and potential explosion in the piping network. On a larger scale, addition of a contaminant or noncompatible chemical substance to the chemical processing system through an unprotected valve could result in the catastrophic release of that contaminant to the general population.

Different security products are available to protect aboveground vs. below ground valves. For example, valve lockout devices can purchased to protect valves and valve controls located aboveground. Vaults containing underground valves can be locked to prevent access to these valves.

As described above, a lockout device can be used as a security measure to prevent unauthorized access to aboveground valves located within petrochemical processing systems. **Valve lockout devices** are locks that are specially designed to fit over valves and valve handles to control their ability to be turned or seated. These devices can be used to lock the valve into the desired position. Once the valve is locked, it cannot be turned unless the locking device is unlocked or removed by an authorized individual.

Various valve lockout options are available for industrial use, including:

- Cable lockouts
- Padlocked chains/cables
- Valve-specific lockouts

Many of these lockout devices are not specifically designed for use in DIB sector industry (i.e., chains, padlocks), but are available from a local hardware store or manufacturer specializing in safety equipment. Other lockout devices (for example valve-specific lockouts or valve box-locks) are more specialized and must be purchased from safety or valve-related equipment vendors.

The three most common types of valves for which lockout devices are available are gate, ball, and butterfly valves. Each is described in more detail below.

- Gate Valve Lockouts—designed to fit over the operating hand wheel of the gate valve to prevent it from being turned. The lockout is secured in place with a padlock. Two types of gate valve lockouts are available: diameter-specific and adjustable. Diameter-specific lockouts are available for handles ranging from 1 inch to 13 inches in diameter. Adjustable gate valve lockouts can be adjusted to fit any handle ranging from 1 inch to 6+ inches in diameter.
- Ball Valve Lockouts—designed to prevent rotation of the valve handle. The three major configurations available are a wedge shape for 1-inch to 3-inch valves, a lockout that completely covers ⅜-inch to 8-inch ball valve handles, and a universal lockout that can be applied to quarter-turn valves of varying sizes and geometric handle dimensions. All three types of ball valve lockouts can be installed by sliding the lockout device over the ball valve handle and securing it with a padlock.

- Butterfly Valve Lockouts—functions in a similar manner to the ball valve lockout. The polypropylene lockout device is placed over the valve handle and secured with a padlock. This type of lockout has been commonly used in the bottling industry.

A major difference between valve-specific lockout devices and the padlocked chain or cable lockouts is that they do not need to be secured to an anchoring device in the floor or the piping system. In addition, valve-specific lockouts eliminate potential tripping or access hazards that may be caused by chains or cable lockouts applied to valves located near walkways or frequently maintained equipment.

Valve-specific lockout devices are available in a variety of colors, which can be useful in distinguishing different valves. For example, different-colored lockouts can be used to distinguish the type of liquid passing through the valve (i.e., treated, untreated, potable, petrochemical), or to identify the party responsible for maintaining the lockout. Implementing a system of different-colored locks on operating valves can increase system security by reducing the likelihood of an operator inadvertently opening the wrong valve and causing a problem in the system.

Security for Vents

Vents are installed in some aboveground DIB sector storage areas to allow safe venting of off-gases. The specific vent design for any given application will vary depending on the design of the chemical storage vessel. However, every vent consists of an open air connection between the storage container and the outside environment. Although these air exchange vents are an integral part of covered or underground chemical storage containers, they also represent a potential security threat. Improving vent security by making the vents tamper-resistant or by adding other security features, such as security screens or security covers, can enhance the security of the entire petrochemical processing system.

Many municipalities already have specifications for vent security at their local chemical industrial assets. These specifications typically include the following requirements:

- Vent openings are to be angled down or shielded to minimize the entrance of surface and/or rainwater into the vent through the opening.
- Vent designs are to include features to exclude insects, birds, animals, and dust.
- Corrosion-resistant materials are to be used to construct the vents.

Visual Surveillance Monitoring

Visual surveillance is used to detect threats through continuous observation of important or vulnerable areas of an asset. The observations can also be recorded for later review or use (for example, in court proceedings). Visual surveillance system can be used to monitor various entry or access points into specific buildings and perimeters of a facility. These systems are also useful in recording individuals who enter or leave a facility, thereby helping to identify unauthorized access. Images can be transmitted live to a monitoring station, where they can be monitored in real time, or they can be recorded and reviewed later. Many DIB sector facilities have found that a combination of electronic surveillance and security guards provides an effective means of facility security.

Visual surveillance is provided through a closed-circuit television (CCTV) system, in which the capture, transmission, and reception of an image is localized within a closed "circuit." This is different than other broadcast images, such as over-the-air television, which is broadcast over the air to any receiver within range.

At a minimum, a CCTV system consists of:

- One or more cameras
- A monitor for viewing the images
- A system for transmitting the images from the camera to the monitor.

Specific attributes and features of camera systems, lenses, and lighting systems are presented in table 9.11.

Table 9.11. Attributes of Camera, Lenses, and Lighting Systems

Attribute	*Discussion*
	Camera Systems
Camera type	Major factors in choosing the correct camera are the resolution of the image required and lighting of the area to be viewed. • **Solid State** (including charge coupled devices, charge priming device, charge injection device, and metal oxide substrate)—these cameras are becoming predominant in the marketplace because of their high resolution and their elimination of problems inherent in tube cameras. • **Thermal**—These cameras are designed for night vision. They require no light and use differences in temperature between objects in the field of view to produce a video image. Resolution is low compared to other cameras, and the technology is currently expensive relative to other technologies. • **Tube**—These cameras can provide high-resolution burn out and must be replaced after one to two years. In addition, tube performance can degrade over time. Finally, tube cameras are prone to burn images in the tube replacement.
Resolution (the ability to see fine details)	User must determine the amount of resolution required depending on the level of detail required for threat determination. A high definition focus with a wide field of vision will give an optimal viewing area.
Field of vision width	Cameras are designed to cover a defined field of vision, which is usually defined in degrees. The wider the field of vision, the more area a camera will be able to monitor.
Type of image produced (color, black-and-white, thermal)	Color images may allow the identification of distinctive markings, while black-and-white images may provide sharper contrast. Thermal imaging allows the identification of heat sources (such as human beings or other living creatures) from low-light environments; however, thermal images are not effective in identifying specific individuals (i.e., for subsequent legal processes).
Pan/tilt/zoom (PTZ)	Panning (moving the camera in a horizontal plane), tilting (moving the camera in a vertical plane), and zooming (moving the lens to focus on objects that are at different distances from the camera) allow the camera to follow a moving object. Different systems allow these functions to be controlled manually or automatically. Factors to be considered in PTZ cameras are the degree of coverage for pan and tilt functions and the power of the zoom lens.
	Lenses
Format	Lens format determines the maximum image size to be transmitted.
Focal length	This is the distance from the lens to the center of the focus. The greater the focal length, the higher the magnification, but the narrower the field of vision.
F number	F number is the ability to gather light. Smaller F numbers may be required for outdoor applications where light cannot be controlled as easily.
Distance and width approximation	The distance and width approximations are used to determine the geometry of the space that can be monitored at the best resolution.

(*continued*)

Table 9.11. ***Continued***

Attribute	*Discussion*
	Lighting Systems
Intensity	Light intensity must be great enough for the camera type to produce sharp images. Light can be generated from natural or artificial sources. Artificial sources can be controlled to produce the amount and distribution of light required for a given camera and lens.
Evenness	Light must be distributed evenly over the field of view so that there are no darker or shadowy areas. If there are lighter vs. darker areas, brighter areas may appear washed out (i.e., details cannot be distinguished) while no specific objects can be viewed from darker areas.
Location	Light sources must be located above the camera so that light does not shine directly into the camera.

USEPA (2005).

MONITORING DEVICES

Proper security preparation really comes down to a three-legged approach: delay, respond, and detect. The third leg of security, to detect, is discussed in this section.

Sensors for Monitoring Petrochemical and Radiological Contamination

Chemical contaminant sensors that can be used to identify potential threats to DIB sector facilities include inorganic monitors (e.g., chlorine analyzer), organic monitors (e.g., total organic carbon analyzer), and toxicity meters. Radiological meters can be used to measure concentrations of several different radioactive species.

Monitoring can be conducted using either portable or fixed-location sensors. Fixed-location sensors are usually used as part of a continuous, on-line monitoring system. Continuous monitoring has the advantage of enabling immediate notification when there is an upset. However, the sampling points are fixed and only certain points in the system can be monitored. In addition, the number of monitoring locations needed to capture the physical, chemical, and biological complexity of a system can be prohibitive. The use of portable sensors can overcome this problem of monitoring many points in the system. Portable sensors can be used to analyze grab samples at any point in the system, but have the disadvantage that they provide measurements only at one point in time.

Radiation Detection Equipment

Radioactive substances (radionuclides) are known health hazards that emit energetic waves and/or particles that can cause both carcinogenic and non-carcinogenic health effects. Radionuclides pose unique threats to source water supplies and petrochemical processing storage, or distribution systems because radiation emitted from radionuclides can affect individuals through several pathways—by direct contact with, ingestion or inhalation of, or external exposure to contaminants. While radiation can occur naturally in some cases due to the decay of some minerals, intentional and non-intentional releases of man-made radionuclides into the environment is also a realistic threat.

Threats to DIB sector and associated petrochemical facilities from radioactive contamination could involve two major scenarios. First, the facility or its assets could be contaminated, preventing personnel from accessing and operating the facility/assets. Second, the feed water supply could be contaminated. These two scenarios require different threat reduction strategies. The first scenario requires that facilities monitor for radioactive substances being brought on-site; the second requires that air and water assets be monitored for radioactive contamination. While the effects of radioactive contamination are basically the same under both threat types, each of these threats requires different types of radiation monitoring and different types of equipment.

COMMUNICATION INTEGRATION

In this section, I discuss those devices necessary for communication and integration of DIB sector industrial processing operations, such as electronic controllers, two-way radios, and wireless data communications. With regard to security applications, electronic controllers are used to automatically activate equipment (such as lights, surveillance cameras, audible alarms, or locks) when they are triggered. Triggering could be in response to a variety of scenarios, including tripping of an alarm or a motion sensor; breaking of a window or a glass door; variation in vibration sensor readings; or simply through input from a timer.

Two-way wireless radios allow two or more users that have their radios tuned to the same frequency to communicate instantaneously with each other without the radios being physically lined together with wires or cables.

Wireless data communications devices are used to enable transmission of data between computer systems and/or between a SCADA server and its sensing devices, without individual components being physically linked together via wires or cables. In DIB sector systems, these devices are often used to link remote monitoring stations (i.e., SCADA components) or portable computers (i.e., laptops) to computer networks without using physical wiring connections.

Electronic Controllers

An electronic controller is a piece of electronic equipment that receives incoming electric signals and uses preprogrammed logic to generate electronic output signals based on the incoming signals. While electronic controllers can be implemented for any application that involves inputs and outputs (for example, control of a piece of machinery in a factor), in a security application, these controllers essentially act as the system's "brain," and can respond to specific security-related inputs with preprogrammed output response. These systems combine the control of electronic circuitry with a logic function such that circuits are opened and closed (and thus equipment is turned on and off) through some preprogrammed logic. The basic principle behind the operation of an electrical controller is that it receives electronic inputs from sensors or any device generating an electrical signal (for example, electrical signals from motion sensors), and then uses it preprogrammed logic to produce electrical outputs (for example, these outputs could turn on power to a surveillance camera or to an audible alarm). Thus, these systems automatically generate a preprogrammed, logical response to a preprogrammed input scenario.

The three major types of electronic controllers are timers, electromechanical relays, and programmable logic controllers (PLCs), which are often called "digital relays." I discuss each of these types of controller in more detail below.

Timers use internal signal/inputs (in contrast to externally generated inputs) and generate electronic output signals at certain times. More specifically, timers control electric current flow to any application to which they are connected, and can turn the current on or off on a schedule pre-specificed by the user. Typical timer range (amount of time that can be programmed to elapse before the timer activates linked equipment) is from 0.2 seconds to 10 hours, although some of the more advanced timers have ranges of up to 60 hours. Timers are useful in fixed applications that don't require frequent schedule changes. For example, a timer can be used to turn on the lights in a room or building at a certain time every day. Timers are usually connected to their own power supply (usually 120–240 V).

In contrast to timers, which have internal triggers based on a regular schedule, electromechanical relays and PLCs have both external inputs and external outputs. However, PLCs are more flexible and more powerful than are electromechanical relays, and thus this section focuses primarily on PLCs as the predominant technology for security-related electronic control applications.

Electromechanical relays are simple devices that use a magnetic field to control a switch. Voltage applied to the relay's input coil creates a magnetic field, which attracts an internal metal switch. This causes the relay's contacts to touch, closing the switch and completing the electrical circuit. This activates any linked equipment. These types of systems are often used for high-voltage applications, such as in some automotive and other manufacturing processes.

Two-Way Radios

Two way radios, as discussed here, are limited to a direct unit-to-unit radio communication, either via single unit-to-unit transmission and reception, or via multiple handheld units to a base station radio contact and distribution system. Radio frequency spectrum limitations apply to all handheld units, as directed by the FCC. This also distinguishes a handheld unit from a base station or base station unit (such as those used by an amateur [ham] radio operator), which operate under different wave length parameters.

Two-way radios allow a user to contact another user or group of users instantly on the same frequency, and to transmit voice or data without the need for wires. They use "half-duplex" communications: communication that can be only transmitted or received—it can't transmit and receive simultaneously. In other words, only one person may talk, while other personnel with radio(s) can only listen. To talk, the user depresses the talk button and speaks into the radio. The audio then transmits the voice wirelessly to the receiving radios. When the speaker has finished speaking and the channel has cleared, users on any of the receiving radios can transmit; either to answer the first transmission or to begin a new conversation. In addition to carrying voice data, many types of wireless radios also allow the transmission of digital data, and these radios may be interfaced with computer networks that can use or track these data. For example, some two-way radios can send information such as global positioning system (GPS) data or the ID of the radio. Some two-way radios can also send data through a SCADA system.

Wireless radios broadcast these voice or data communications over the airwaves from the transmitter to the receiver. While this can be an advantage in that the signal emanates in all directions and does not need a direct physical connection to be received at the receiver, it can also make the communications vulnerable to being blocked, intercepted, or otherwise altered. However, security features are available to ensure that the communications are not tampered with.

Wireless Data Communications

A wireless data communication system consists of two components: a "wireless access point" (WAP), and a "wireless network interface card" (sometimes also referred to as a "client"), which work together to complete the communications link. These wireless systems can link electronic devices, computers, and computer systems together using radio waves, thus eliminating the need for these individual components to be directly connected together through physical wires. Wireless connections do have a couple of limitations. First, wireless data connections are limited by the distance between components (radio waves scatter over a long distance and cannot be received efficiently, unless special directional antenna are used). Second, these devices function only if the individual components are in a direct line of sight with each other, since radio waves are affected by interference from physical obstructions. However, in some cases, repeater units can be used to amplify and retransmit wireless signals to circumvent these problems. The two components of wireless devices—WAP and wireless network interface card/client—are discussed in more detail below.

The **WAP** provides the wireless data communication service. It usually consists of a housing (which is constructed from plastic or metal depending on the environment it will be used in) containing a circuit board; flash memory that holds software; one of two external ports to connect to existing wired networks; a wireless radio transmitter/receiver; and one or more antenna connections. Typically, the WAP requires a one-time user configuration to allow the device to interact with the local area network (LAN). This configuration is usually done via a web-driven software application which is accessed via a computer.

Wireless Network Interface Card/Client: A wireless card is a piece of hardware that is plugged into a computer and enables that computer to make a wireless network connection. The card consists of a transmitter, functional circuitry, and a receiver for the wireless signal, all of which work together to enable communication between the computer, its wireless transmitter/receiver, and its antenna connection. Wireless cards are installed in a computer through a variety of connections, including USB adapters, or laptop vardBus (PCMCIA) or desktop peripheral (PCI) cards. As with the WAP, software is loaded onto the user's computer, allowing configuration of the card so that it may operate over the wireless network.

Two of the primary applications for wireless data communications systems are to enable mobile or remote connections to a LAN, and to establish wireless communications links between SCADA remote telemetry units (RTUs) and sensors in the field. Wireless car connections are usually used for LAN access from mobile computers. Wireless cards can also be incorporated into RTUs to allow them to communicate with sensing devices that are located remotely.

CYBER PROTECTION DEVICES

Various cyber protection devices are currently available for use in protecting DIB sector computer systems. These protection devices include anti-virus and pest eradication software, firewalls, and network intrusion hardware/software. These products are discussed in this section.

Anti-Virus and Pest Eradication Software

Anti-virus programs are designed to detect, delay, and respond to programs or pieces of code that are specifically designed to harm computers. These programs are known as **malware**. Malware can include computer viruses, worms, and Trojan horse programs (programs that appear to be benign but which have hidden harmful effects).

Pest eradication tools are designed to detect, delay, and respond to **spyware** (strategies that websites use to track user behavior, such as by sending "cookies" to the user's computer), and hacker tools that track keystrokes (keystroke loggers) or passwords (password crackers).

Viruses and pests can enter a computer system through the Internet or through infected USBs (thumb drives) or CDs. They can also be placed into a system by insiders. Some of these programs, such as viruses and worms, then move within a computer's drives and files, or between computers if the computers are networked to each other. This malware can deliberately damage files, utilize memory and network capacity, crash application programs, and initiate transmissions of sensitive information from a PC. While the specific mechanisms of these programs differ, they can infect files and even the basic operating program of the computer firmware/hardware.

The most important features of an anti-virus program are its abilities to identify potential malware and to alert a user before infection occurs, as well as its ability to respond to a virus already resident on a system. Most of these programs provide a log so that the user can see what viruses have been detected and where they were detected. After detecting a virus, the anti-virus software may delete the virus automatically, or it may prompt the user to delete the virus. Some programs will also fix files or programs damaged by the virus.

Various sources of information are available to inform the general public and computer system operators about new viruses being detected. Since anti-virus programs use signatures (or snippets of code or data) to detect the presence of a virus, periodic updates are required to identify new threats. Many anti-virus software providers offer free upgrades that are able to detect and respond to the latest viruses.

Firewalls

A **firewall** is an electronic barrier designed to keep computer hackers, intruders, or insiders from accessing specific data files and information on a DIB sector's computer network or other electronic/computer systems. Firewalls operated by evaluating and then filtering information coming through a public network (such as the Internet) into the utility's computer or other electronic system. This evaluation can include identifying the source or destination addresses and ports, and allowing or denying access based on this identification.

There are two methods used by firewalls to limit access to the utility's computers or other electronic systems from the public network:

- The firewall may deny all traffic unless it meets certain criteria, or
- The firewall may allow all traffic through unless it meets certain criteria.

A simple example of the first method is to screen requests to ensure that they come from an acceptable (i.e., previously identified) domain name and Internet protocol address. Firewalls may also use more complex rules that analyze the application data to determine if the traffic should be allowed through. For example, the firewall may require user authentication (i.e., use of a password) to access the system. How a firewall determines what traffic to let through depends on which network layer it operates at and how it is configured. Some of the pros and cons of various methods to control traffic flowing in and out of the network are provided in table 9.12.

Table 9.12. Pros and Con of Various Firewall Methods for Controlling Network Access

Method	*Description*	*Pros*	*Cons*
Packet filtering	Incoming and outgoing packets (small chunks of data) are analyzed against a set of filters. Packets that make it through the filters are sent to the requesting system and all others are discarded. There are two type of packet filtering: static (the most common) and dynamic.	Static filtering is relatively inexpensive, and little maintenance required. It is well-suited for closed environments where access to or from multiple addresses is not allowed.	Leaves permanent open holes in the network; allows direct connection to internal hosts by external sources; offers no user authentication, method can be unmanageable in large networks.
Proxy service	Information from the Internet is retrieved by the firewall and then sent to the requesting system and vice versa. In this way, the firewall can limit the information made known to the requesting system, making vulnerabilities less apparent.	Only allows temporary open holes in the network perimeter. Can be used for all types of internal protocol services.	Allows direct connections to internal hosts by external clients; offers no user authentication

(*continued*)

Table 9.12. *Continued*

Method	*Description*	*Pros*	*Cons*
Stateful pattern recognition	This method examines and compares the contents of certain key parts of an information packet against a database of acceptable information. Information traveling from inside the firewall to the outside is monitored for specific defining characteristics, then incoming information is compared to these characteristics. If the comparison yield a reasonable match, the information is allowed through. If not, the information is discarded.	Provides a limited time window to allow pockets of information to be sent; does not allow any direct connections between internal and external hosts; supports user-level authentication.	Slower than packet filtering; does not support all types of connections.

USEPA (2005).

Firewalls may be a piece of hardware, a software program, or an appliance card that contains both. Advanced features that can be incorporated into firewalls allow for the tracking of attempts to log on to the local area network system. For example, a report of successful and unsuccessful log-in attempts may be generated for the computer specialist to analyze. For systems with mobile users, firewalls allow remote access in to the private network by the use of secure log-on procedures and authentication certificates. Most firewalls have a graphical user interface for managing the firewall.

In addition, new Ethernet firewall cards that fit in the slot of an individual computer bundle additional layers of defense (like encryption and permit/deny) for individual computer transmissions to the network interface function. These new cards have only a slightly higher cost than traditional network interface cards.

Network Intrusion Hardware/Software

Network intrusion detection and prevention system are software- and hardware-based programs designed to detect unauthorized attacks on a computer network system.

While other applications, such as firewalls and anti-virus software, share similar objectives with network intrusion systems, network intrusion systems provide a deeper layer of protection beyond the capabilities of these other systems because they evaluate pattern of computer activity rather than specific files.

It is worth noting that attacks may come from either outside or within the system (i.e., from an insider), and that network intrusion detection systems may be more applicable for detecting patterns of suspicious activity from inside a facility (accessing sensitive data, etc.) than are other information technology solutions.

Network intrusion detection systems employ a variety of mechanisms to evaluate potential threats. The types of search and detection mechanisms are dependent upon the level of sophistication of the system. Some of the available detection methods include:

- Protocol analysis—the process of capturing, decoding, and interpreting electronic traffic. The protocol analysis method of network intrusion detection involves the analysis of data captured during transactions between two or more systems or devices, and the evaluation of these data to identify unusual activity and potential problems. Once a problem is isolated and recorded, problems or potential threats can be linked to pieces of hardware or software. Sophisticated protocol analysis will also provide statistics and trend information on the captured traffic.
- Traffic anomaly detection—identifies potential threatening activity by comparing incoming traffic to "normal" traffic patterns and identifying deviations. It does this by comparing user characteristics against thresholds and triggers defined by the network administrator. This method is designed to detect attacks that span a number of connections, rather than a single session.
- Network honeypot—establishes non-existent services in order to identify potential hackers. A network honeypot impersonates services that don't exist by sending fake information to people scanning the network. It identifies

the attacker when they attempt to connect to the service. There is no reason for legitimate traffic to access these resources because they don't exist; therefore any attempt to access them constitutes an attack.

- Anti-intrusion-detection system evasion techniques—designed against attackers who may be trying to evade intrusion-detection system scanning. They include methods called IP defragmentation, TCP streams reassembly, and deobfuscation.

While these detection systems are automated, they can indicate only patterns of activity, and a computer administer or other experienced individual must interpret activities to determine whether or not they are potentially harmful. Monitoring the logs generated by these systems can be time consuming, and there may be a learning curve to determine a baseline of "normal" traffic patterns from which to distinguish potential suspicious activity.

REFERENCES AND RECOMMENDED READING

Garcia, M.L. (2001). *The Design and Evaluation of Physical Protection Systems.* Butterworth-Heinemann.

IBWA (2004). *Bottled Water Safety and Security.* Alexandria, VA: International Bottled Water Association.

NAERC (2002). *Security Guidelines for the Electricity Sector.* Washington, DC: North American Electric Reliability Council.

Schneier, B. (2000). *Secrets and Lies: Digital Security in a Networked World.* New York: Wiley.

USEPA (2005). *Water and Wastewater Security Product Guide.* Accessed 4//4/09 at https://cfpub.epa.gov/safewater/watersecurity/guide. (No longer available on the USEPA website.)

10

Fourteen Features of Active and Effective Security

> New York is an international symbol of freedom and democracy. That's what we are. And we are proud of it. That also makes us a target for those people who oppose those concepts.
>
> —Andrew Cuomo, Governor of New York, October 31, 2017

The events of 9/11 dramatically changed this nation and focused us on combating terrorism. As a result, in 2003 and subsequent years, the Department of Homeland Security (DHS) in conjunction with members from the general public, state and local agencies, and private groups concerned with the safety of critical infrastructures, established a Water Security Working Group (WSWG) to consider and make recommendations on infrastructure security issues. Although initially created to make recommendations for water/wastewater security, WSWG is an excellent template for use with other critical infrastructures, including DIB assets. For example, the WSWG identified active and effective security practices for critical infrastructure, and provide an approach for adopting these practices. It also recommended mechanisms to provide incentives that facilitate broad and receptive response among critical infrastructure sectors to implement active and effective security practices. Finally, WSWG recommended mechanisms to measure progress and achievements in implementing active and effective security practices, and identify barriers to implementation.

The WSWG recommendations on security are structured to maximize benefits to critical industries by emphasizing actions that have the potential both to improve the quality or reliability of service, and to enhance security. These recommendations, based on original recommendations from the 2003 National Drinking Water Advisor Council (NDWAC), were designed primarily, as the name suggests, for use by water systems of all types and sizes, including systems that serve less than 3,300 people. However, it is the author's opinion, based on personal experience, that NDWAC's recommendations, when properly adapted to applicable circumstances and locations, can be applied to any and all critical infrastructure sectors, including the DIB sector.

The NDWAC identified fourteen features of active and effective security programs that are important to increasing security and relevant across the broad range of utility circumstances and operating conditions. USEPA (2003) points out that the fourteen features are, in many cases, consistent with the steps needed to maintain technical, management, and operational performance capacity related to overall water quality; as pointed out earlier, these steps can be applied to other critical infrastructures as well. Many facilities may be able to adopt some of the features with minimal, if any, capital investment.

FOURTEEN FEATURES OF ACTIVE AND EFFECTIVE SECURITY

It is important to point out that the fourteen features of active and effective programs emphasize that "one size does not fit all" and is not a cookie-cutter approach to effective implementation of security measures. There will be variability in security approaches and tactics among DIB sector facilities, based on industry-specific circumstances and operating conditions. The fourteen features:

- Are sufficiently flexible to apply to all DIB industries, regardless of size
- Incorporate the idea that active and effective security programs should have measurable goals and time lines
- Allow flexibility for DIB sector industrial facilities to develop specific security approaches and tactics that are appropriate to industry-specific circumstances

DIB sector facilities can differ in many ways including:

- Transportation supply source (rail, air, water, pipeline, lines, or ground)
- Number of supply sources

- Energy processing capacity
- Operation risk
- Location risk
- Security budget
- Spending priorities
- Political and public support
- Legal barriers
- Public vs. private ownership

DIB sector industrial facilities should address security in an informed and systematic way, regardless of these differences. DIB sector facilities need to fully understand the specific, local circumstances and conditions under which they operate, and develop a security program tailored to those conditions. The goal in identifying common features of active and effective security programs is to achieve consistency in security program outcomes among DIB sector industrial facilities, while allowing for and encouraging facilities to develop utility-specific security approaches and tactics. The features are based on a comprehensive "security management layering system" approach that incorporates a combination of public involvement and awareness, partnerships, and physical, chemical, operational, and design controls to increase overall program performance. They address industry security in four functional categories: organization, operation, infrastructure, and external. These functional categories are discussed in greater detail below.

- **Organizational**—There is always something that can be done to improve security. Even when resources are limited, the simple act of increasing organizational attentiveness to security may reduce vulnerability and increase responsiveness. Preparedness itself can help deter attacks. The first step to achieving preparedness is to make security a part of the organizational culture, so that it is in the day-to-day thinking of front-line employees, emergency responders, and management of every DIB sector facility in this country. To successfully incorporate security into "business as usual," there must be a strong commitment to security by organization leadership and by the supervising body, such as the board of stockholders. The following features address how a security culture can be incorporated into an organization.
- **Operational**—In addition to having a strong culture and awareness of security within an organization, an active and effective security program

makes security part of operational activities, from daily operations, such as monitoring of physical access controls, to scheduled annual reassessments. DIB sector industries will often find that by implementing security into operations they can also reap cost benefits, and improve the quality or reliability of the energy service.

- **Infrastructure**—These recommendations advise utilities to address security in all elements of DIB sector industry infrastructure—from source to distribution and through processing and product delivery.
- **External**—Strong relationships with response partners and the public strengthen security and public confidence. Two of the recommended features of active end effective security programs address this need.

The Fourteen Features

Feature 1. Make an explicit and visible commitment of the senior leadership to security.

DIB sector industrial facilities should create an explicit, easily communicated, enterprise-wide commitment to security, which can be done through:

- Incorporating security into a utility-wide mission or vision statement, addressing the full scope of an active and effective security program—that is, protection of worker/public health, worker/public safety, and public confidence, and that is part of core day-to-day operations
- Developing an enterprise-wide security policy, or set of policies

DIB sector industries should use the process of making a commitment to security as an opportunity to raise awareness of security throughout the organization, making the commitment visible to all employees and customers, and to help very facet of the enterprise to recognize the contribution they can make to enhancing security.

Feature 2. Promote security awareness throughout the organization.

The objective of a security culture should be to make security awareness a normal, accepted, and routine part of day-to-day operations. Examples of tangible efforts include:

- Conducting employee training
- Incorporating security into job descriptions

- Establishing performance standards and evaluations for security
- Creating and maintaining a security tip line and suggestion box for employees
- Making security a routine part of staff meetings and organization planning
- Create a security policy

Feature 3. Assess vulnerabilities and periodically review and update vulnerability assessments to reflect changes in potential threats and vulnerabilities.

Because circumstances change, DIB sector industrial facilities should maintain their understanding and assessment of vulnerabilities as a "living document," and continually adjust their security enhancement and maintenance priorities. DIB sector industrial facilities should consider their individual circumstances and establish and implement a schedule for review of their vulnerabilities.

Assessments should take place once every three to five years at a minimum. DIB sector industries may be well served by doing assessments annually.

The basic elements of sound vulnerability assessments are:

- Characterization of the chemical processing system, including its mission and objectives
- Identification and prioritization of adverse consequences to avoid
- Determination of critical assets that might be subject to malevolent acts that could result in undesired consequences
- Assessment of the likelihood (qualitative probability) of such malevolent acts from adversaries
- Evaluation of existing countermeasures
- Analysis of current risk and development of a prioritized plan for risk reduction

Feature 4. Identify security priorities, and, on an annual basis, identify the resources dedicated to security programs and planned security improvements, if any.

Dedicated resources are important to ensure a sustained focus on security. Investment in security should be reasonable considering utilities' specific circumstances. In some circumstances, investment may be as simple as increasing the amount of time and attention that executives and managers give to security. Where threat potential or potential consequences are greater, greater investment likely is warranted.

This feature establishes the expectation that chemical industrial facilities should, through their annual capital, operations, maintenance, and staff resources plans, identify and set aside resources consistent with their specific identified security needs. Security priorities should be clearly documented and should be reviewed with utility executives at least once per year as part of the traditional budgeting process.

Feature 5. Identify managers and employees who are responsible for security and establish security expectations for all staff.

- Explicit identification of security responsibilities is important for development of a security culture with accountability.
- At minimum, DIB sector industrial facilities should identify a single, designated individual responsible for overall security, even if other security roles and responsibilities will likely be dispersed throughout the organization.
- The number and depth of security-related roles will depend on a utility's specific circumstances.

Feature 6. Establish physical and procedural controls to restrict access to chemical industrial infrastructure to only those conducting authorized, official business and to detect unauthorized physical intrusions.

Examples of physical access controls include fencing critical areas, locking gates and doors, and installing barriers at site access points. Monitoring for physical intrusion can include maintaining well-lighted facility perimeters, installing motion detectors, and utilizing intrusion alarms. The use of neighborhood watches, regular employee rounds, and arrangements with local police and fire departments can support identifying unusual activity in the vicinity of facilities.

Examples of procedural access controls include, inventorying keys, changing access codes regularly, and requiring security passes to pass gates an access sensitive area. In addition, utilities should establish the means to readily identify all employees including contractors and temporary workers with unescorted access to facilities.

Feature 7. Employee protocols for detection of contamination consistent with the recognized limitations in current contaminant detection, monitoring, and surveillance technology.

Until progress can be made in development of practical and affordable online contaminant monitoring and surveillance systems, most DIB sector industrial facilities must use other approaches to contaminant monitoring and surveillance.

Many utilities already measure the above parameters (and many others) on a regular basis to control plant operations and confirm chemical mixture quality; more closely monitoring these parameters may create operational benefits for facilities that extend far beyond security, such as reducing operating costs and chemical usage. DIB sector industrial facilities also should thoughtfully monitor customer complaints and improve connections with local public health networks to detect public health anomalies. Customer complaints and public health anomalies are an important way to detect potential contamination problems and other environmental quality concerns.

Feature 8. Define security-sensitive information; establish physical, electronic, and procedural controls to restrict access to security-sensitive information; detect unauthorized access; and ensure information and communications systems will function during emergency response and recover.

Protecting IT systems largely involves using physical hardening and procedural steps to limit the number of individuals with authorized access and to prevent access by unauthorized individuals. Examples of physical steps to harden SCADA and IT networks include installing and maintaining firewalls and screening the network for viruses. Examples of procedural steps include restricting remote access to data networks, and safeguarding critical data through backups and storage in safe places. Utilities should strive for continuous operation of IT and telecommunications systems, even in the event of an attack, by providing uninterruptible power supply and backup systems, such as satellite phones.

In addition to protecting IT systems, security sensitive information should be identified and restricted to the appropriate personnel. Security sensitive information could be contained within:

- Facility maps and blueprints
- Operations details
- Hazardous material utilization and storage
- Tactical-level security program details

- Any other information on utility operations or technical details that could aid in planning or execution of an attack

Identification of security-sensitive information should consider all ways that utilities might use and make public information (e.g., many chemical industrial facilities may at times engage in competitive bidding processes for construction of new facilities or infrastructure). Finally, information critical to the continuity of day-to-day operations should be identified and backed up.

Feature 9. Incorporate security considerations into decisions about acquisition, repair, major maintenance, and replacement of physical infrastructure; include consideration of opportunities to reduce risk through physical hardening and adoption of inherently lower-risk design and technology options.

Prevention is a key aspect of enhancing security. Consequently, consideration of security issues should begin as early as possible in facility construction (i.e., it should be a factor in building plans and designs). However, to incorporate security considerations into design choices, chemical facilities need information about the types of security design approaches and equipment that are available and the performance of these designs and equipment in multiple dimensions. For example, DIB sector facilities would want to evaluate not just the way that a particular design might contribute to security, but would also look at how that design would affect the efficiency of day-to-day plant operations and worker safety.

Feature 10. Monitor available threat-level information, and escalate security procedures in response to relevant threats.

Monitoring threat information should be a regular part of a security program manager's job, and utility-, facility-, and region-specific threat levels and information should be shared with those responsible for security. As part of security planning, DIB sector facilities should develop systems to access threat information, procedures that will be followed in the event of increased industry or facility threat levels, and should be prepared to put these procedures in place immediately, so that adjustments are seamless. Involving local law enforcement and FBI is critical.

DIB sector facilities should investigate what networks and information sources might be available to them locally, and at the state and regional level. If a utility cannot gain access to some information networks, attempts should be made to align with those who can and will provide effective information to the DIB sector facility.

Feature 11. Incorporate security considerations into emergency response and recovery plans, test and review plans regularly, and update plans to reflect changes in potential threats, physical infrastructure, chemical processing operations, critical interdependencies, and response protocols in partner organizations.

DIB sector facilities should maintain response and recovery plans as "living documents." In incorporating security considerations into their emergency response and recovery plans, chemical facilities also should be aware of the National Incident Management System (NIMS) guidelines, established by DHS, and of regional and local incident management commands and systems, which tend to flow from the national guidelines.

DIB sector facilities should consider their individual circumstances and establish, develop, and implement a schedule for review of emergency response and recovery plans. DIB sector facility plans should be thoroughly coordinated with emergency response and recovery planning in the larger community. As part of this coordination, a mutual and program should be established to arrange in advance for exchanging resources (personnel or physical assets) among agencies within a region, in the event of an emergency or disaster that disrupts operation. Typically, the exchange of resource is based on a written formal mutual and agreement. For example, Florida's Water-Wastewater Agency Response Network (FlaWARN), deployed after Hurricane Katrina, and allowed the new "utilities helping utilities" network to respond to urgent requests from Mississippi for help to bring facilities back on line after the hurricane.

The emergency response and recovery plans should be reviewed and updated as needed annually. This feature also establishes the expectation that chemical facilities should test or exercise their emergency response and recovery plans regularly.

Feature 12. Develop and implement strategies for regular, ongoing security-related communications with employees, response organizations, rate setting organizations, and customers.

An active and effective security program should address protection of public health, public safety (including infrastructure), and public confidence. DIB sector facilities should create an awareness of security and an understanding of the rationale for their overall security management approach in the communities they reside in and/or serve.

Effective communication strategies consider key messages; who is best equipped/trusted to deliver the key messages; the need for message consistency, particularly during an emergency; and the best mechanisms for delivering messages and for receiving information and feedback from key partners. The key audiences for communication strategies are utility employees, response organizations, and customers.

Feature 13. Forge reliable and collaborative partnerships with the communities served, managers of critical interdependent infrastructure, response organizations, and other local utilities.

Effective partnerships build collaborative working relationships and clearly define roles and responsibilities, so that people can work together seamlessly if an emergency should occur. It is important for DIB sector facilities within a region and neighboring regions to collaborate and establish a mutual aid program with neighboring utilities, response organizations, and sectors, such as the power sector, on which utilities rely or impact. Mutual aid agreements provide for help from other organizations that is prearranged and can be accessed quickly and efficiently in the event of a terrorist attack or natural disaster. Developing reliable and collaborative partnerships involves reaching out to managers and key staff and other organizations to build reciprocal understanding and to share information about the facility's security concerns and planning. Such efforts will maximize the efficiency and effectiveness of a mutual aid program during an emergency response effort, as the organizations will be familiar with each other's circumstances, and thus will be better able to serve each other.

It is also important for DIB sector facilities to develop partnerships with the communities and customers they serve. Partnerships help to build credibility within communities and establish public confidence in utility operations. People who live near DIB sector facility structures can be the eyes and ears of the facility, and can be encouraged to notice and report changes in operating procedures or other suspicious behaviors.

DIB sector facilities and public health organizations should establish formal agreements on coordination to ensure regular exchange of information between facilities and public health organizations, and outline roles and responsibilities during response to and recovery from an emergency. Coordination is important at all levels of the public health community—national public health, county health agencies, and health care providers, such as hospitals.

Feature 14. Develop DIB facility-specific measures of security activities and achievements, and self-assess against these measures to understand and document program progress.

Although security approaches and tactics will be different depending on chemical utility-specific circumstances and operating conditions, we recommend that all DIB sector facilities monitor and measure a number of common types of activities and achievements, including existence of program policies and procedures, training, testing, and implementing schedules and plans.

The Fourteen Feature Matrix

In the following table, a matrix of recommended measures to assess effectiveness of a DIB sector facility's security program is presented. Each feature is grouped according to its functional category: organization, operation, infrastructure, and external.

Table 10.1. Fourteen Features of Active and Effective Security Matrix

Features	*Checklist: Potential Measures of Progress*
Organizational Features	
Feature 1—Explicit commitment to security	Does a written, enterprise-wide security policy exist, and is the policy reviewed regularly and updated as needed?
Feature 2—Promote security awareness	Are incidents reported in a timely way, and are lessons learned from incident responses reviewed and, as appropriate, incorporated into future utility security efforts?
Feature 5—Defined security roles and employee expectations	Are managers and employees who are responsible for security identified?

(continued)

Table 10.1. ***Continued***

Features	*Checklist: Potential Measures of Progress*
	Operational Features
Feature 3—Vulnerability assessment up to date	Are reassessments of vulnerabilities made after incidents, and are lessons learned and other relevant information incorporated into security practices?
Feature 4—Security resources and implementation priorities	Are security priorities clearly identified, and to what extent do security priorities have resources assigned to them?
Feature 7—Contamination detection	Is there a protocol/procedure in place to identify and respond to suspected contamination events?
Feature 10—Threat-level based protocols	Is there a protocol/procedure of responses that will be made if threat levels change?
Feature 11—Emergency response plan tested and up to date	Do exercise address the full range of threats—physical, cyber, and contamination—and is there a protocol/procedure to incorporate lessons learned from exercises and actual response into updates to emergency response and recovery plans?
Feature 14—Industry-specific measures and self-assessment	Does the utility perform self-assessment a least annually?
	Infrastructure Features
Feature 6—Intrusion detection and access control	To what extent are methods to control access to sensitive assets in place?
Feature 8—Information protection and continuity	Is there a procedure to identify and control security-sensitive information, is information correctly categorized, and how to control measures perform under testing?
Feature 9—Design and construction standards	Are security considerations incorporated into internal utility design and construction standards for new facilities/infrastructure and major maintenance projects?
	External Features
Feature 12—Communications	Is there a mechanism for utility employees, partners and the community to notify the utility of suspicious occurrences and other security concerns?
Feature 13—Partnerships	Have reliable and collaborative partnerships with customers, managers of independent interrelated infrastructure, and response organizations been established?

USEPA (2003).

Ultimately, the goal of implementing the fourteen security features (and all other security provisions) is to create a significant improvement in DIB sector industry on a national scale, by reducing vulnerabilities, and therefore risk to public health from terrorist attacks and natural disasters. To create a sustainable effect, the DIB sector as a whole must not only adopt and actively practice the features, but also incorporate the features into "business as usual."

REFERENCES AND RECOMMENDED READING

USEPA (2003). *Active and Effective Water Security Programs.* Accessed 6/06 at https://cfpub.epa.gov/safewater/watersecurity/14 features.cfm. (No longer available on the USEPA website.)

Appendix

Definition of Terms

Access control—Any combination of barriers, gates, electronic security equipment, and/or guards that can deny entry to unauthorized personnel or vehicles.

Access control point—A station at entrance to a building or a portion of a building where identification is checked and people and hand-carried items are searched.

Acid bomb—A crude bomb made by combining muriatic acid with aluminum strips in a two-liter soda bottle.

Active vehicle barrier—An impediment placed at an access control point that may be manually or automatically deployed in response to detection of a threat.

Aerosol—A fine mist or spray, which contains minute particles suspended in a gas (e.g., fog, smoke).

Agency—A division of government with a specific function, or a non-governmental organization (e.g., private contractor, business) that offers a particular kind of assistance. In the incident command system, agencies are defined as jurisdictional (having statutory responsibility for incident mitigation) or assisting and/or cooperating (providing resources and/or assistance).

Aggressor—Any person seeking to compromise a function or structure.

Airborne contamination—Chemical or biological agents introduced into and fouling the source of supply of breathing or conditioning air.

Airlock—A building entry configuration with which airflow from the outside can be prevented from entering a toxic-free area. An airlock uses two doors, only one of which can be opened at a time, and a blower system to maintain positive air pressures and purge contaminated air from the airlock before the second door is opened.

Al Qaeda—Meaning "the Base"; an international terrorist group founded in approximately 1989 and dedicated to opposing non-Islamic governments with force and violence. One of the principal goals of al Qaeda was to drive the U.S. armed forces out of the Saudi Arabian peninsula and Somalia by violence. Currently wanted for several terrorist attacks, including those on the U.S. embassy in Kenya and Tanzania as well as the first and second World Trade Center bombings, and the attack on the Pentagon.

Al Tahwid—A Palestinian group based in London which professes a desire to destroy both Israel and the Jewish people throughout Europe. Eleven al Tahwid were arrested in Germany allegedly as they were about to begin attacking that country.

Alarm assessment—Verification and evaluation of an alarm alert through the use of closed-circuit television (CCTV) or human observation. Systems used for alarm assessment are designed to respond rapidly, automatically, and predictably to the receipt of alarms at the security center.

Alarm priority—A hierarchy of alarms by order of importance. This is often used in larger systems to give priority to alarms with greater importance.

All-Hazards—A grouping classification encompassing all conditions, environmental or man-made, that have the potential to cause injury, illness, or death; damage to or loss of equipment, infrastructure services, or property, or alternatively causing functional degradation to social, economic, or environmental aspects.

Alpha radiation—The least penetrating type of nuclear radiation. Not considered dangerous unless particles enter the body.

American Airlines Flight 11—The Boeing 767 carrying eighty-one passengers, nine flight attendants, and two pilots, which was hijacked and crashed into the north tower of the World Trade Center at 8:45 a.m. Eastern Time, on September 11, 2001. Flight 11 was en route to Los Angeles from Boston.

American Airlines Flight 77—The Boeing 757 carrying fifty-eight passengers, four flight attendants, and two pilots, which was hijacked and crashed into the Pentagon at 9:40 a.m. Eastern Time, on September 11, 2001. Flight was en route to Los Angeles from Dulles International Airport in Virginia.

Ammonium nitrate-fuel oil (ANFO)—A powerful explosive made by mixing fertilizer and fuel oil. The type of bomb used in the first World Trade Center attack as well as the 1995 Oklahoma City bombing.

Annunciation—A visual, audible, or other indication by a security system of a condition.

Anthrax—An often fatal infectious disease contracted from animals. Anthrax spores have such a long survival period; the incubation period is short; disability is severe, making anthrax a bioweapon of choice by several nations.

Antidote—A remedy to counteract the effects of poison.

Antigen—A substance which stimulates an immune response by the body immune system recognizes such substances as foreign and produces antibodies to fight them.

Antiterrorism—Defensive measures used to reduce the vulnerability of individuals, forces, and property to terrorist acts.

Antitoxin—An antibody which neutralizes a biological toxin.

Area lighting—Lighting that illuminates a large exterior area.

Assessment—The evaluation and interpretation of measurements and other information to provide a basis for decision-making.

Asset—A person, structure, facility, information, material or process that has value. In the context of the National Infrastructure Protection Plan (NIPP), people are not considered assets.

Asymmetric threat—The use of crude or low-tech methods to attack a superior or more high-tech enemy.

Axis of Evil—Iran, Iraq, and North Korea as defined by President George W. Bush during his State of the Union speech in 2002 as nations that were a threat to U.S security due to harboring terrorism.

Audible alarm device—An alarm device that produces an audible announcement (e.g., bell, horn, siren) of an alarm condition.

Balanced magnetic switch—A door position switch using a reed switch held in a balanced or center position by interacting magnetic fields when not in alarm condition.

Ballistics attack—An attack in which small arms (e.g., pistols, submachine guns, shotguns, rifles) are fired from a distance and rely on the flight of the projectile to damage the target.

Bioaccumulative—Substances that concentrate in living organisms as the breather contaminated air drink or live in contaminated water or eat contaminated food rather than being eliminated through natural processes.

Biochemical warfare—Collective term for use of both chemical warfare and biological warfare weapons.

Biochemterroism—Terrorism using as weapons biological or chemical agents.

Biological agents—Living organisms or the materials derived from them that cause disease in or harm to humans, animals, or plants or cause deterioration of material. Biological agents may be used as liquid droplets, aerosols, or dry powders.

Biological ammunition—Ammunition designed specifically to release a biological agent used as the warhead for biological weapons. Biological ammunition may take many forms, such as a missile warhead or bomb.

Biological attacks—The deliberate release of germs or other biological substances that cause illness.

Biometric reader—A device that gathers and analyzes biometric features.

Biometrics—The use of physical characteristics of the human body as a unique identification method.

Biosafety Level 1—Suitable for work involving well-characterized biological agents not known to consistently cause disease in healthy adult humans, and of minimal potential hazard to lab personnel and the environment. Work is generally conducted on open bench tops using standard microbiological practices.

Biosafety Level 2—Suitable for work involving biological agents of moderate potential hazard to personnel and the environment. Lab personnel should have specific training in handling pathogenic agents and be directed by competent scientists. Access to the lab should be limited when work is being conducted, extreme precautions should be taken with contaminated sharp items, and certain procedures should be conducted in biological safety cabinets or other physical containment equipment if there is a risk of creating infectious aerosols or splashes.

Biosafety Level 3—Suitable for work done with indigenous or exotic biological agents that may cause serious or potentially lethal disease as a result of exposure by inhalation. Lab personnel must have specific training in handling pathogenic and potentially lethal agents and be supervised by competent scientists who are experienced in working with these agents. All procedures involving the manipulation of infectious material are conducted within biological safety cabinets or other physical containment devices, or by personnel wearing appropriate personal protective clothing and equipment. The lab must have special engineering and design features.

Biosafety Level 4—Suitable for work with the most infectious biological agents. Access to the two Biosafety Level 4 labs in the United States is highly restricted.

Bioterrorism—The use of biological agents in a terrorist operation. Biological toxin would include anthrax, Ricin, botulism, the plague, smallpox, and tularemia.

Bioterrorism Act—The Public Health Security and Bioterrorism Preparedness and Response Act of 2002.

Biowarfare—The use of biological agents to cause harm to targeted people either directly, by bringing the people into contact with the agents or indirectly, by infecting other animals and plants, which would in turn cause harm to the people.

Blast curtains—Heavy curtains made of blast-resistant materials that could protect the occupants of a room from flying debris.

Blast-resistant glazing—Window opening glazing that is resistant to blast effects because of the interrelated function of the frame and glazing material properties frequently dependent upon tempered glass, polycarbonate, or laminated glazing.

Blast vulnerability envelope—The geographical area in which an explosive device will cause damage to assets.

Blister agents—Agents which cause pain and incapacitation instead of death and might be used to injure many people at once, thereby overloading medical facilities and causing fear in the population. Mustard gas is the best known blister agent.

Blood agents—Agents based on cyanide compounds. More likely to be used for assassination than for terrorism.

Bollard—A vehicle barrier consisting of a cylinder, usually made of steel and sometimes filled with concrete, placed on end in the ground and spaced about three feet apart to prevent vehicles from passing, but allowing entrance of pedestrians and bicycles.

Botulism—An illness caused by the botulinum toxin, which is exceedingly lethal and quite simple to produce. It takes just a small amount of the toxin to destroy the central nervous system. Botulism may be contracted by the ingestion of contaminated food or through breaks or cuts in the skin. Food supply contamination or aerosol dissemination of the botulinum toxin are the two ways most likely to be used by terrorists.

Boundary penetration sensor—An interior intrusion-detection sensor that detects attempts by individuals to penetrate or enter a building.
Building—A structure, usually enclosed by walls and a roof, constructed to provide support or shelter for an intended occupancy. Note that other structures, such as canopies, are not considered buildings for the purposes of security standards.
Building hardening—Enhanced construction that reduces vulnerability to external blast and ballistic attacks.
Building separation—The distance between closest points on the exterior walls of adjacent buildings or structures.
Bush Doctrine—The policy that holds responsible nations which harbor or support terrorist organizations and says that such countries are considered hostile to the U.S. From President George W. Bush's speech: "A country that harbors terrorists will either deliver the terrorist or share in their fate. . . . People have to choose sides. They are either with the terrorists, or they're with us."
Business continuity—The ability of an organization to continue to function before, during, and after a disaster.
BWC—Officially known as the "Convention on the Prohibition of Development, Production, and Stockpiling of Bacteriological (Biological) and Toxin Weapons and Destruction." The BWC works toward general and complete disarmament, including the prohibition and elimination of all types of weapons of mass destruction.
Cable barrier—Cable or wire rope anchored to and suspended off the ground or attached to chain-link fence to act as a barrier to moving vehicles.
Camp X-Ray—The Guantanamo Bay, Cuba, prison that houses al Qaeda and Taliban prisoners.
Capacitance sensor—A device that detects an intruder approaching or touching a metal object by sensing a change in capacitance between the object and the ground.
Card reader—A device that gathers or reads information when a card is presented as an identification method.
Carrier—A person or animal that is potentially a source of infection by carrying on infectious agent without visible symptoms of the disease.
Cascading event—The occurrence of one event that causes another event.
Causative agent—The pathogen, chemical, or other substance that is the cause of disease or death in an individual.

Cell—The smallest unit within a guerrilla or terrorist group. A cell generally consists of two to five people dedicated to a terrorist cause. The formation of cells is born of the concept that an apparent "leaderless resistance" makes it hard for counterterrorists to penetrate.

Chain of custody—The tracking and documentation of physical control of evidence.

Chemical agent—A toxic substance intended to be used for operations to debilitate, immobilize, or kill military or civilian personnel.

Chemical ammunition—A munition, commonly a missile, bomb, rocket, or artillery shell, designed to deliver chemical agents.

Chemical attack—The intentional release of toxic liquid, gas or solid in order to poison the environment or people.

Chemical warfare—The use of toxic chemicals as weapons, not including herbicide used to defoliate battlegrounds or riot control agents such as gas or mace.

Chemical weapons—Weapons that produce effects on living targets via toxic chemical properties. Examples would be sarin, VX nerve gas, or mustard gas.

Chemterrorism—The use of chemical agents in a terrorist operation. Well-known chemical agents include sarin and VX nerve gas.

Chimney effect—Air movement in a building between floors caused by differential air temperature (differences in density) between the air inside and outside the building. It occurs in vertical shafts, such as elevators, stairwells, and conduit/wiring/piping chases. Hotter air inside the building will rise and he replaced by infiltration with colder outside air through the lower portions of the building. Conversely, reversing the temperature will reverse the flow (down the chimney). Also known as stack effect.

Choking agent—Compounds that injure primarily in the respiratory tract (i.e., nose, throat, and lungs). In extreme cases membranes swell up, lungs become filled with liquid, and death results from lack of oxygen.

Cipro—A Bayer antibiotic that combats inhalation anthrax.

Clear zone—An area that is clear of visual obstructions and landscape materials that could conceal a threat or perpetrator.

Closed-circuit television (CCTV)—An electronic system of cameras, control equipment, recorders, and related apparatus used for surveillance or alarm assessment.

Collateral damage—Injury or damage to assets that are not the primary target of an attack.

Community—A political entity that has the authority to adopt and enforce laws and ordinances for the area under its jurisdiction. In most cases, the community is an incorporated town, city, township, village, or unincorporated area of a country; however, each state defines its own political subdivisions and forms of government.

Components and cladding—Elements of the building envelope that do not qualify as part of the main wind-force resisting system.

Consequence analysis—The estimate of the potential public health and economic impacts that a successful attack could cause.

Consequence management—Measures to protect public health and safety, restore essential governmental services, and provide emergency relief to governments, businesses, and individuals affected by the consequences of terrorism. State and local governments exercise the primary authority to respond to the consequences of terrorism.

Contamination—The undesirable deposition of a chemical, biological, or radiological (CBR) material on the surface of structures, areas, objects, or people.

Continuity of services and operations—Controls to ensure that, when unexpected events occur, departmental/agency minimum essential infrastructure services and operations, including computer operations, continue without interruption or are promptly resumed, and that critical and sensitive data are protected through adequate contingency and business recovery plans and exercise.

Control center—A sophisticated monitoring and control system responsible for balancing power generation and demand; monitoring flows over transmission lines to avoid overloading; planning and configuring the system to operate reliably; maintaining system stability; preparing for emergencies; and placing equipment in and out of service for maintenance and emergencies.

Control systems—Computer-based systems used within many infrastructures and industries to monitor and control sensitive processes and physical functions. These systems typically collect measurement and operation data from the field, process and display the information, and relay control commands to local or remote equipment or human-machine interfaces (operators). Examples of types of control systems include SCADA system, process control systems, and distributed control systems.

Controlled area—An area into which access is controlled or limited. The portion of a restricted area usually near or surrounding a limited or exclusion area. Correlates with exclusion zone.

Controlled lighting—Illumination of specific areas or sections.

Controlled perimeter—A physical boundary at which vehicle and personnel access is controlled at the perimeter of a site. Access control at a controlled perimeter should demonstrate the capacity to search individuals and vehicles.

Conventional construction—Building construction that is not specifically designed to resist weapons, explosives, or CBR effects. Conventional construction is designed only to resist common loadings and environmental effects such as wind, seismic, and snow loads.

Counterintelligence—Information gathered and activities conducted to protect against: espionage, other intelligence activities, sabotage, or assassinations conducted for or on behalf of foreign powers, organizations, or persons; or international terrorist activities, excluding personnel, physical, document, and communications security programs.

Counterterrorism—Measures used to prevent preempt, or retaliate against terrorist attacks.

Covert entry—Attempts to enter a facility by using false credentials or stealth.

Crash bar—A mechanical egress device located on the interior side of a door that unlocks the door when pressure is applied in the direction of egress.

Crime Prevention through Environmental Design (CPTED)—A crime prevention strategy based on evidence that the design and form of the built environment can influence human behavior. CPTED usually involves the use of three principles: natural surveillance (by placing physical features, activities, and people to maximize visibility); natural access control (through the judicial placement of entrances, exits, fencing, landscaping, and lighting); and territorial reinforcement (using buildings, fences, pavement, signs, and landscaping to express ownership).

Crisis management—The measures taken to identify, acquire, and plant the use of resources needed to anticipate, prevent, and/or resolve a threat or act of terrorism.

Critical assets—Those assets essential to the minimum operations of the organization, and to ensure the health and safety of the general public.

Critical infrastructure—Systems and assets, whether physical or virtual, so vital that the incapacity or destruction of such may have a debilitating impact on the security, economy, public health or safety, environment, or any combination of these matters, across any federal. State, regional, territorial, or local jurisdiction.

Criticality—A description of the importance of a particular sector asset, system, network, or function in relation to national or regional security issues. Includes a consideration of public health and economic impacts.

Cutaneous—Related to or entering through the skin.

Cutaneous anthrax—Anthrax that is contracted via broken skin. The infection spreads through the bloodstream causing cyanosis, shock, sweating, and finally death.

Cyanide agent—Used by Iraq in the Iran war against the Kurds in the 1980s, and also by the Nazis in the gas chambers of concentration camps, cyanide agents are colorless liquid which is inhaled in its gaseous form while liquid cyanide and cyanide salts are absorbed by the skin. Symptoms are headache, palpitations, dizziness, and respiratory problems followed later by vomiting, convulsions, respiratory failure and unconsciousness, and eventually by death.

Cybersecurity—The prevention of damage to, unauthorized use of, or exploitation of, and, if needed, the restoration of electronic information and communications systems and the information contained therein to ensure confidentially, integrity, and availability. Includes protection and restoration, when needed, or information networks and wireline, wireless, satellite, public safety answering points, and 911 communications systems and control systems.

Cyber system—Any combination of facilities, equipment, personnel, procedures, and communications integrated to provide cyber services. Examples include business systems, control systems, and access control systems.

Cyberterrorism—Attacks on computer networks or systems, generally by hackers working with or for terrorist groups. Some forms of cyberterrorism include denial of service attacks, inserting viruses or stealing data.

Data transmission equipment—A path for transmitting data between two or more components (e.g., a sensor and alarm reporting system, a card reader and controller, a CCTV camera and monitor, a transmitter and receiver).

Decontamination—The reduction or removal of a CBR material from the surface of a structure, area, object, or person.

Defensive layer—Building design or exterior perimeter barriers intended to delay attempted forced entry.

Defensive measures—Protective measures that delay or prevent attack on an asset or that shield the asset from weapons, explosives, and CBR effects. Defensive measures include site work and building design.

Delay rating—A measure of the effectiveness of penetration protection of a defense layer.

Dependency—The one-directional reliance of an asset, system network, or collection thereof, within or across sectors, in input, interaction, or other requirement from other sources in order to function properly.

Design basis threat—The threat (e.g., tactics and associated weapons, tools, explosives) against which assets within a building must be protected and upon which the security engineering design of the building is based.

Design constraint—Anything that restricts the design options for a protective system or that creates additional problems for which the design must compensate.

Design opportunity—Anything that enhances protection, reduces requirements for protective measures, or solves a design problem.

Detection layer—A ring of intrusion-detection sensors located on or adjacent to a defensive layer or between two defensive layers.

Detection measures—Protective measures that detect intruders, weapons, or explosives; assist in assessing the validity of detection; control access to protected areas; and communicate the appropriate information to the responsive force. Detection measures include detection systems, assessment systems, and access control system elements.

Detection system elements—Detection measures that detect the presence of intruders, weapons, or explosives. Detection system elements include intrusion-detection systems, weapons and explosives detectors, and guards.

Dirty bomb—A makeshift nuclear device which is created from radioactive nuclear waste material. While not a nuclear blast, an explosion of a dirty bomb causes localized radioactive contamination as the nuclear waste material is carried into the atmosphere where it is dispersed by the wind.

Disaster—An occurrence of a natural catastrophe, technological accident, or human-caused event that has resulted in severe property damage, deaths, and/or multiple injuries.

Domestic terrorism—The unlawful use, or threatened use, or force or violence by a group or individual based and operating entirely within the United States or Puerto Rico without foreign direction committed against persons or property to intimidate or coerce a government, the civilian population, or any segment thereof in furtherance of political or social objectives.

Door position switch—A switch that changes state based on whether or not a door is closed. Typically, a switch mounted in a frame that is actuated by a magnet in a door.

Door strike, electronic—An electromechanical lock that releases a door plunger to unlock the door. Typically, an electronic door strike is mounted in place or near a normal door strike plate.

Dose rate (radiation)—A general term indicating the quantity (total or accumulated) or ionizing radiation or energy absorbed by a person or animal, per unit of time.

Dosimeter—An instrument for measuring and registering total accumulated exposure to ionizing radiation.

Dual technology sensor—A sensor that combines two different technologies in one unit.

Duress alarm devices—Also known as panic buttons, these devices are designated specifically to initiate a panic alarm.

eBomb (or e-bomb)—Electromagnetic bomb which produces a brief pulse of energy which affects electronic circuitry. At low levels, the pulse temporarily disables electronics systems, including computers, radios, and transportation systems. High levels completely destroy circuitry, causing mass disruption of infrastructure while sparing life and property.

Ecotage—Is the portmanteau of the "eco-" prefix and "sabotage." It is used to describe illegal acts of vandalism and violence, committed in the name of environmental protection.

Ecoterrorism—A neologism for terrorism that includes sabotage intended to hinder activities that are considered damaging to the environment.

Effective standoff distance—A standoff distance at which the required level of protection can be shown to be achieved through analysis or can be achieved through building hardening or other mitigating constitution or retrofit.

Electromagnetic Pulse (EMP)—A burst of electromagnetic radiation by deliberate means, such as nuclear attack, or through natural means, such as a large-scale geomagnetic storm. Magnetic and electric fields resulting from EMP have the potential to disrupt electrical and electronic systems by causing destructive current and voltage surges.

Electronic emanations—Electromagnetic emissions from computers, communications, electronics wiring, and related equipment.

Electronic emanations eavesdropping—Use of electronic-emanation surveillance equipment from outside a facility or its restricted area to monitor electronic emanations from computers, communications, and related equipment.

Electronic entry control systems—Electronic devices that automatically verify authorization for a person to enter or exit a controlled area.

Electronic security system—An integrated system that encompasses interior and exterior sensors, CCTV systems for assessment of alarm conditions, electronic entry control systems, data transmission media, and alarm reporting systems for monitoring, control, and display of various alarm and system information.

Emergency alert system—A communications system of broadcast stations and interconnecting facilities authorized by the Federal Communications Commission (FCC). The system provides the president and other federal, state and local officials the means to broadcast emergency information to the public before, during, and after disasters.

Entry control point—A continuously or intermittently manned station at which entry to sensitive or restricted areas is controlled.

Entry control stations—Entry control stations should be provided at main perimeter entrances where security personnel are present. Entry control stations should be located as close as practical to the perimeter entrance to permit personnel inside the station to maintain constant surveillance over the entrance and its approaches.

Evacuation—Organized, phased, and supervised dispersal of people from dangerous or potentially dangerous areas.

Exclusion area—A restricted area containing a security interest. Uncontrolled movement permits direct access to the item.

Facial recognition—A biometric technology that is based on features of the human face.

Fallout—The descent to the earth's surface of particles contaminated with radioactive material from a radioactive cloud. The term can also be applied to the contaminated particulate matter itself.

Fatah—Meaning "conquest by means of jihad"; a political organization created in the 1960s and led by Yasser Arafat. With both a military and intelligence wing, it has carried out terrorist attacks on Israel since 1965. It joined the PLO in 1968. Since 9/11, the Fatah was blamed for attempting to smuggle fifty tons of weapons into Israel.

Fence protection—An intrusion-detection technology that detects a person crossing a fence by various methods such as climbing, crawling, cutting, and so on.

Fence sensor—An exterior intrusion-detection sensor that detects aggressors as they attempt to climb over, cut through or otherwise disturbs a fence.

Fenestration—A building opening.

Fiber optics—A method of data transfer by passing bursts of light through a strand of glass or clear plastic.

Field of view—The visible area in a video picture.

Filtrate—In ultrafiltration, the water that passes through the membrane and which contains particles smaller than the molecular weight cutoff of the membrane.

Forced entry—Entry to a denied area achieved through force to create an opening in fence, walls, doors, etc., or to overpower guards.

Fragment-retention film (FRF)—A thin, optically clear film applied to glass to minimize the spread of glass fragments when the glass is shattered.

Frame rate—In digital video, a measurement of the rate of change in a series of pictures, often measured in frames per second.

Frangible construction—Building components that are designed to fail to vent blast pressures from an enclosure in a controlled manner and direction.

Frustration-aggression hypothesis—A hypothesis that every frustration leads to some form of aggression and every aggressive act results from some prior frustration.

Function—A service, process, capability, or operation performed by an asset, system, network, or organization.

Fundamentalism—Conservative religious authoritarianism. Fundamentalism is not specific to Islam; it exists in all faiths. Characteristics include literal interpretation of scriptures and a strict adherence to traditional doctrines and practices.

Geneva Protocol 1925—The first treaty to prohibit the use of biological weapons. The 1925 Geneva Protocol for the Prohibition of the Use in War of Asphyxiating, Poisonous, or Other Gases and Bacteriological Methods of Warfare.

Germ warfare—The use of biological agents to cause harm to targeted people either directly, by bringing the people into contact with the agents or indirectly, by infecting other animals and plants, which would in turn cause harm to the people.

Glanders—An infectious bacterial disease known to cause inflammation in horses, donkeys, mules, goats, dogs and cats. Human infection has not been seen since 1945, but because so few organisms are required to cause disease, it is considered a potential agent for biological warfare.

Glare security lighting—Illumination projected from a security perimeter into the surrounding area, making it possible to see potential intruders at a considerable distance while making it difficult to observe activities within the secure perimeter.

Glass-break detector—An intrusion-detection sensor that is designed to detect breaking glass either through vibration or acoustics.

Glazing—A material installed in a sash, ventilator, or panes (e.g., glass, plastic, thin granite installed in a curtain wall).

Government Coordinating Council (GCC)—The government counterpart to the SCC for each sector established to enable interagency coordination. The GCC comprises representatives across various levels of government (federal, state, local, tribal, and territorial) as appropriate to the security and operations landscape of each individual sector.

Grab sample—A single sample collected at a particular time and place that represents the composition of the water, air, or soil only at that time and location.

Grid wire sensor—An intrusion-detection sensor that uses a grid of wires to cover a wall or fence. An alarm is sounded if the wires are cut.

Ground zero—From 1946 until 9/11, ground zero was the point directly above, below, or at which a nuclear explosion occurs or the center or origin of rapid, intense, or violent activity or change. After 9/11, the term, when used with initial capital letters, refers to the ground at the epicenter of the World Trade Center attacks.

Guerrilla warfare—The term was invented to describe the tactics Spain used to resist Napoleon, though the tactic itself has been around much longer. Literally, it means "little war." Guerilla warfare features cells and utilizes no front line. The oldest form of asymmetric warfare, guerilla warfare is based on sabotage and ambush with the objective of destabilizing the government through lengthy and low-intensity confrontation.

Hacktivist—One who hacks into computer networks to achieve a certain political aim—sometimes only to create disruption to direct attention to a specific cause.

Hamas—A radical Islamic organization which operates primarily in the West Bank and Gaza Strip whose goal is to establish an Islamic Palestinian state in place of Israel. On the one hand, Hamas operates overtly in their capacity as social services deliverers, but its activists have also conducted many attacks, including suicide bombings, against Israeli civilians and military targets.

Hand geometry—A biometric technology that is based on characteristics of the human hand.

Hardening—See *Building hardening.*

Hazard—An inherent physical or chemical characteristic that has the potential for causing harm to people, the environment, or property.

Hazard assessment—The process of evaluating available information about the site to identify potential hazards that might pose a risk to the site characterization team. The hazard assessment results in assigning one of four levels to risk: lower hazard, radiological hazard, high chemical hazard, or high biological hazard.

Hazard mitigation—Any action taken to reduce or eliminate the long-term risk to human life and property from hazards. The term is sometimes used in a stricter sense to mean cost-effective measures to reduce the potential for damage to a facility or facilities form a disaster event.

Hazardous material (HazMat)—Any substance or material that, when involved in an accident and released in sufficient quantities, poses a risk to people's health, safety, and/or property. These substances and materials include explosives, radioactive materials, flammable liquids or solids, combustible liquids or solids, poisons, oxidizers, toxins, and corrosive materials.

Hemorrhagic fevers—In general, the term viral hemorrhagic fever is used to describe severe multisystem syndrome wherein the overall vascular system is damaged, and the body becomes unable to regulate itself. These symptoms are often accompanied by hemorrhage; however, the bleeding itself is not usually life-threatening. Some types of hemorrhagic fever viruses can cause relatively mild illnesses.

High-hazard areas—Geographic locations that, for planning purposes, have been determined through historical experience and vulnerability analysis to be likely to experience the effects of a specific hazard (e.g., hurricane, earthquake, hazardous materials accident), resulting in vast property damage and loss of life.

High-Impact, Low-Frequency (HILF)—HILF evens are occurrences that are relatively unusual, but have the potential to cause catastrophic disruption. Examples include pandemic disease, terrorist attack, and electromagnet pulse.

High-risk target—Any material resource or facility that, because of mission sensitivity, ease of access, isolation, and symbolic value, may be an especially attractive or accessible terrorist target.

Hizbollah (Hezbollah)—Meaning "The Party of God." One of many terrorist organizations which seek the destruction of Israel and of the United States. They have taken credit for numerous bombings against civilians and have declared that civilian targets are warranted. Hezbollah claims it sees no legitimacy for the existence of Israel, and that their conflict becomes one of legitimacy that is based on religious ideals.

Homeland Security Office—An agency organized after 9/11, with former Pennsylvania Governor Tom Ridge heading it up. The Office of Homeland Security is at the top of approximately 40 federal agencies charged with protecting the United States against terrorism.

Homicide bombings—A term the White House coined the term to replace the old "suicide bombings."

Human-caused hazard—Human-caused hazards are technological hazards and terrorism. They are distinct from natural hazards primarily in that they originated from human activity. Within the military services, the term threat is typically used for human-caused hazard.

Impact analysis—A management level analysis that identifies the impacts of losing the entity's resources. The analysis measures the effect of resource loss and escalating losses over time in order to provide the entity with reliable data upon which to base decisions on hazard mitigation and continuity planning.

Improvised explosive device (IED)—A device placed or fabricated in an improvised manner incorporating destructive, lethal, noxious, pyrotechnic, or incendiary chemicals and designed to destroy, incapacitate, harass, or distract. It may incorporate conventional military construction (e.g., artillery round), but is normally devised form nonmilitary components.

Incident—A confirmed occurrence that requires response actions to prevent or minimize loss of life or damage to property and/or natural resources. A drinking water contamination incident occurs when the presence of a harmful contaminant has been confirmed.

Infrastructure—The framework of interdependent networks and systems comprising identifiable industries, (including people and procedures), and distribution capabilities that provide a reliable flow of products and services essential to the defense of economic security of the United States, the smooth functioning of government at all levels, and society as a whole. Consistent with the definition in the Homeland Security Act, infrastructure includes physical, cyber, and/or human elements.
Inhalation anthrax—A form of anthrax that is contracted by inhaling anthrax spores. This results in pneumonia, sometimes meningitis, and finally death.
Insider compromise—A person authorized access to a facility (an insider) compromises assets by taking advantage of that accessibility.
Insider threat—An aggressor who is an employee of a business, institution, or agency seeking to compromise a function or the building of the employer.
Intercom system—An electronic system that allows simplex, half-duplex, or full-duplex audio communications.
Interdependency—Mutually reliant relationship between entities (objects, individuals, or groups). The degree of interdependency does not need to be equal in both directions.
Integrated rapid vessel screening (IRVS)—Quick and simple procedure to assess the risk and resiliency of a building, mass transit station, or tunnel in an all-hazard (man-made and natural) context.
International terrorism—Violent acts or acts dangerous to human life that are a violation of the criminal laws of the United States or any state, or that would be a criminal violation if committed within the jurisdiction of the United States or any State. These acts appear to be intended to intimidate or coerce a civilian population, influence the policy of a government by intimidation or coercion, or affect the conduct of a government by assassination or kidnapping. International terrorist acts occur outside the United States, or transcend national boundaries in terms of the means by which they are accomplished, the persons they appear intended to coerce or intimidate, or the locale in which their perpetrators operate or seek asylum.
Intifada (intifadah)—(alternatively Intifadah, from Arabic "shaking off") The two intifadas are similar in that both were originally characterized by civil disobedience by the Palestinians which escalated into the use of terror. In 1987, following the killing of several Arabs in the Gaza Strip, the first intifada

began and went on until 1993. The second intifada began in September 2000, following Ariel Sharon's visit to the Temple Mount.

Intrusion-detection sensor—A device that initiates alarm signals by sensing the stimulus, change, or condition for which it was designed.

Intrusion-detection system—The combination of components, including sensors, control units, transmission lines, and monitor units, integrated to operate in a specified manner.

ISIS (ISIL) Islamic State of Iraq and the Levant—Islamic extremist rebel group that presently controls territory and brutalizes and/or murders various groups in various countries, including Syria, Iraq, Libya, and Nigeria, and others.

Isolated fenced perimeters—Fenced perimeters with 100 feet or more of space outside the fence that is clear of obstruction, making approach obvious.

Jersey barrier—A protective concrete barrier initially and still used as a highway divider that now also functions as an expedient method for traffic speed control at entrance gates and to keep vehicles away from buildings.

Jihad—Meaning "struggle." The definition is a subject of vast debate. There are two definitions generally accepted. The first is a struggle against oppression, whether political or religious. The second is the struggle within oneself, or a spiritual struggle.

Key resources—As defined in the Homeland Security Act, key resources are publicly or privately controlled resources essential to the minimal operations of the economy and government.

LD50—The dose of a substance which kills 50 percent of those infected.

Laboratory Response Network (LRN)—A network of labs developed by the CDC, APHL, and FBI for the express purpose of dealing with bioterrorism threats, including pathogens and some biotoxins.

Laminated glass—A flat-lite™ of uniform thickness of two monolithic glass plies boned together with an interlayer material as defined in ASTM Specification C1172. Many different interlayer materials are used in laminated glass.

Landscaping—The use of plantings (shrubs and trees), with or without landforms and/or large boulders, to act as a perimeter barrier against defined threats.

Laser card—A card technology that uses a laser reflected off of a card for uniquely identifying the card.

Lassa fever—An acute, often fatal, viral disease characterized by high fever, ulcers of the mucous membranes, headaches, and disturbances of the gastrointestinal system.

Layers of Protection—A traditional approach in security engineering using concentric circles extending out from an area to be protected as demarcation points for different security strategies.

Level of protection—The degree to which an asset is protected against injury or damage from an attack.

Life cycle costs (LCC)—The combined sum of all relevant costs associated with owning and operating a constructed facility over a specificed period of time, usually the estimated life space of a building.

Limited area—A restricted area within close proximity of a security interest. Uncontrolled movement may permit access to the item. Escorts and other internal restrictions may prevent access to the item.

Line of sight—Direct observation between two points with the naked eye or hand-held optics.

Line-of-sight sensor—A pair of devices used as an intrusion-detection sensor that monitor any movement through the field between the sensors.

Links—The means (road, rail, barge, or pipeline) by which a chemical is transported from one node to another.

Magnetic lock—An electromagnetic lock that unlocks a door when power is removed.

Magnetic stripe—A card technology that uses a magnetic stripe on the card to encored data used for unique identification of the card.

Mail-bomb delivery—Bombs or incendiary devices delivered to the target in letters or packages.

Man-trap—An access control strategy that uses a pair of interlocking doors to prevent tailgating. Only one door can be unlocked at a time.

Mass notification—Capability to provide real-time information to all building occupants or personnel in the immediate vicinity of a building during emergency situations.

Microwave motion sensor—An intrusion-detection sensor that uses microwave energy to sense movement within the sensor's field of view. These sensors work similar to radar by using the Doppler effect to measure a shift in frequency.

Mindset—According to *American Heritage Dictionary*: "1. A fixed mental attitude or disposition that predetermines a person's response to an interpretation of situations; 2. and inclination or a habit." *Merriam Webster's Collegiate Dictionary* (10th ed.) defines it as 1. A mental attitude or inclination; 2. a fixed state of mind. The term dates from 1926.

Minimum-efficiency reporting value (MERV)—Commonly known as MERV rating is a measurement scale to rate the effectiveness of air filters. The MERV rating is from 1 to 16. Higher MERV ratings correspond to a greater percentage of particles captured on each pass, with a MERV 16 filter capturing more than 95 percent of particles over the full range. Minimum essential infrastructure resource elements. The broad categories of resources, all or portions of which constitutes the minimal essential infrastructure necessary for a department, agency, or organization to conduct its core mission(s).

Minimum measures—Protective measures that can be applied to all buildings regardless of the indentified threat. These measures offer defense or detection opportunities for minimal cost, facilitate future upgrades, and may deter acts of aggression.

Mitigation—Ongoing and sustained action to reduce the probability of or lessen the impact of an adverse incident.

Molotov cocktail—A crude incendiary bomb made of a bottle filled with flammable liquid and fitted with a rag wick.

Monkeypox—The Russian bioweapon program worked with this virus, which is in the same family as smallpox. In June 2003, a spate of human monkeypox cases was reported in the U.S. Midwest. This was the first time that monkeypox was seen in North America, and it was the first time that monkeypox was transferred from animal to human. There was some speculation that it was a bioattack.

Motion detector—An intrusion-detection sensor that changes state based on movement in the senor's field of view.

Moving vehicle bomb—An explosive-lade car or truck driven into or near a building and detonated.

Mustard gas—Blistering agents which cause severe damage to the eyes, internal organs, and respiratory system. Produced for the first time in 1822, mustard gas was not used until World War I. Victims suffered the effects of mustard gas thirty to forty years after exposure.

Narcoterrorism—The view of many counterterrorist experts that there exists an alliance between drug traffickers and political terrorists.

National Pharmaceutical Stockpile—A stock of vaccines and antidotes which are stored at Centers for Disease Control in Atlanta, to be used against biological warfare.

Natural hazard—Naturally occurring events such as floods, earthquakes, tornadoes, tsunami, coastal storms, landslides, and wildfires that strike populated areas. A natural event is a hazard when it has the potential to harm people or property. The risks of natural hazards may be increased or decreased as a result of human activity; however, they are not inherently human-induced.

Natural protective barriers—Natural protective barriers are mountains and deserts, cliffs and ditches, water obstacles, or other terrain features that are difficult to transverse.

Nerve agent—The Nazis used the first nerve agents: insecticides developed into chemical weapons. Some of the better known nerve agents include VX, sarin, soman, and tabun. These agents are used because only a small quantity is necessary to inflict a substantial damage. Nerve agents can be inhaled or can absorb through intact skin.

Network—A group of components that share information or interact with each other in order to perform a function.

Nodes—A facility at which a chemical is produced, stored, or consumed.

Non-exclusive zone—An area around an asset that has controlled entry, but shared or less restrictive access than an exclusive zone.

Non-persistent agent—An agent that, upon release, loses its ability to cause casualties after ten to fifteen minutes. It has a high evaporation rate, is lighter than air, and will disperse rapidly. It is considered to be a short-term hazard; however, in small, unventilated areas, the agent will be more persistent

Nuclear, biological, or chemical weapons—Also called weapons of mass destruction (WMD). Weapons that are characterized by their capability to produce mass casualties.

Nuclear blast—An explosion of any nuclear material which is accompanied by a pressure wave, intense light and heat, and widespread radioactive fallout which can contaminate the air, water, and ground surface for miles around.

Open systems architecture—A term borrowed from the IT industry to claim that systems are capable of interfacing with other systems from any vendor, which also uses open system architecture. The opposite would be a proprietary system.

Operator interface—The part of a security management system that provides that user interface to humans.

Opportunity Contaminant—A contaminant that might be readily available in a particular area, even though they may not be highly toxic or infectious or easily dispersed and stable in treated drinking water.

Osama bin Laden (also spelled "Usama")—A native of Saudi Arabia, was born the seventeenth of twenty-four sons of Saudi Arabian builder Mohammed bin Oud bin Laden, a Yemeni immigrant. Early in his career, he helped the mujahedeen fight the Soviet Union by recruiting Arabs and building facilities. He hated the United States because he viewed the United States as having desecrated holy ground in Saudi Arabia with their presence during the first Gulf War. Expelled from Saudi Arabia in 1991 and from Sudan in 1996, he operated terrorist training camps in Afghanistan. His global network al Qaeda was credited with the attacks on the United States on September 11, 2001, the attack on the USS *Cole* in 2000, and a number of other terrorist attacks. He was killed by U.S. special forces on May 2, 2011.

Owners/operators—Those entities responsible for day-to-day operation and investment in a particular asset or system.

Pandemic influenza—Defined by the World Health Organization (WHO) as a global outbreak of influenza, characterized by an emergent strain of the virus, little to no immunity among the general population, rapid and sustained person-to-person transmission, and lack of a vaccine. On June 11, 2009, WHO determined that 2009 H1N1 influenza (also known as "swine flu") had reached pandemic status.

Pan-tilt-zoom (PTZ) camera. A camera that can move side to side, up and down, and zoom in or out.

Passive infrared motion sensor—A device that detects a change in the thermal energy pattern caused by a moving intruder and initiates an alarm when the change in energy satisfies the detector's alarm criteria.

Passive vehicle barrier—A vehicle barrier that is permanently deployed and does not require response to be effective.

Patch panel—A concentrated termination point that separates backbone cabling from devices cabling for easy maintenance and troubleshooting.

Pathogen—Any agent which can cause disease.

Pathways—The sequence of nodes and links by which a chemical is produced, transported, and transformed from its initial source to its ultimate consumer.

Performance-based design (PBD)—The process used to achieve performance levels for specific attributes based on quantifiable benchmark metrics that can be verified.

Performance measure—Indicator, statistic, or metric used to gauge program performance.

Perimeter barrier—A fence, wall, vehicle barrier, landform, or line of vegetation applied along an exterior perimeter used to obscure vision, hinder personnel access, or hinder or prevent vehicle access.

Persistent agent—An agent that, upon release, retains its casualty-producing effects for an extended period of time, usually anywhere from thirty minutes to several days. A persistent agent usually has a low evaporation rate and its vapor is heavier than air; therefore, its vapor cloud tends to hug the ground. It is considered to be a long-term hazard. Although inhalation hazardous are still a concern, extreme caution should be taken to avoid skin contact as well.

Physical security—The use of barriers and surveillance to protect resources, personnel, and facilities against crime, damage, or unauthorized access.

Plague—The pneumonic plague, which is more likely to be used in connection with terrorism, is naturally carried by rodents and fleas but can be aerosolized and sprayed from crop dusters. A 1970 World Health Organization assessment asserted that, in a worst case scenario, a dissemination of 50 kg in an aerosol over a city of five million could result in 150,000 cases of pneumonic plague, 80,000–100,000 of which would require hospitalization, and 36,000 of which would be expected to die.

Planter barrier—A passive vehicle barrier, usually constructed of concrete and filled with dirt (and flowers for aesthetics). Planters, along with bollards, are the usual street furniture used to keep vehicles away from existing buildings. Overall size and the depth of installation below grade determine the vehicle stopping cap ability of the individual planter.

Plume—Airborne material spreading from a particular source; the dispersal of particles, gases, vapors, and aerosols into the atmosphere.

Political terrorism—Terrorist acts directed at governments and their agents and motivated by political goals (i.e., national liberation).

Polycarbonate glazing—A plastic glazing material with enhanced resistance to ballistics or blast effects.

Potassium iodide—An FDA-approved nonprescription drug for use as a blocking agent to prevent the thyroid gland from absorbing radioactive iodine.

Predetonation screen—A fence that causes an anti-tank round to detonate or prevents it from arming before it reaches its target.

Preparedness—The activities necessary to build, sustain, and improve readiness capabilities to prevent, protect against, respond to, and recover from natural or man-made incidents. Preparedness is a continuous process involving efforts at all levels of government and between government and the private sector and nongovernmental organizations to identify threats, determine vulnerabilities, and identify required resources to prevent, respond to, and recover from major incidents.

Pressure mat—A mat that generates an alarm when pressure is applied to any part of the mat's surface, such as when someone steps on the mat. Pressure mats can be used to detect an intruder approaching a protected object, or they can be placed by doors and windows to detect entry.

Presumptive results—Results of chemical and/or biological field testing that need to be confirmed by further lab analysis. Typically used in reference to the analysis of pathogens.

Prevention—Actions taken and measures put in place for the continual assessment and readiness of necessary actions to reduce the risk of threats and vulnerabilities, to intervention and stop an occurrence, or to mitigate effects.

Primary asset—An asset that is the ultimate target for compromise by an aggressor.

Prioritization—In the context of the NIPP, prioritization is the process of using risk assessment results to identify where risk reduction or mitigation efforts are most needed and to subsequently determine which protective actions should be instituted in order to have the greatest effect.

Probability of detection—A measure of an intrusion-detection sensor's performance in detecting an intruder within its detection zone.

Probability of intercept—The probability that an act of aggression will be detected and that a response force will intercept the aggressor before the asset can be compromised.

Progressive collapse—A chain reaction failure of building members to an extent disproportionate to the original localized damage.

Protection—Actions or measures taken to cover of shield from exposure, injury, or destruction. In the context of the NIPP, protection includes actions to deter the threat, mitigate the vulnerabilities, or minimize the consequences associated with a terrorist attack or other incident. Protection can include a wide range of activities, such as hardening facilities, building resilience and redundancy, incorporating hazard resistance into initial facility design, initiating active or passive countermeasures, installing security systems and redundancy, incorporating hazard resistant into initial facility design, initiating active or passive countermeasures, installing security systems, promoting workforce surety, training and exercises, and implementing cybersecurity measures, among various others.

Protective barriers—Define the physical limits of a site, activity, or area by restricting, channeling, or impeding access and forming a continuous obstacle around the object.

Protective measures—Elements of a protective system that protect an asset against a threat. Protective measures are divided not defensive and detection measures.

Protective system—An integration of all of the protective measures required to protect an asset against the range of threats applicable to the asset.

Proximity sensor—An intrusion-detection sensor that changes stated based on the close distance or contact of a human to the sensor. These sensors often measure the change in capacitance as a human body enters the measured field.

Psychopathy—A mental disorder, especially an extreme mental disorder marked usually by egocentric and antisocial activity, according to *Webster's*.

Psychopathology—The study of psychological and behavioral dysfunction occurring in mental disorder or in social disorganization, according to *Webster's*.

Psychotic—Of, relating to, or affected with psychosis, which is a fundamental mental derangement (as schizophrenia) characterized by defective or lost contact with reality, according to *Webster's*.

Radiation—High-energy particles or gamma rays that are emitted by an atom as the substance undergoes radioactive decay. Particles can be either charged alpha or beta particles or neutral neutron or gamma rays.

Radiation sickness—The symptoms characterizing the sickness known as radiation injury, resulting from excessive exposure of the whole body to ionizing radiation.

Radio frequency (RF) data transmission—A communications link using RF to send or receive data.

Radiological monitoring—The process of locating and measuring radiation by means of survey instruments that can detect and measure (as exposure rates) ionizing radiation.

Rapid field testing—Analysis of water during site characterization uses rapid field water testing technology in an attempt to tentatively identify contaminants or unusual water quality.

Recovery—The development, coordination, and execution of service- and site-restoration plans for affected communities and the reconstitution of government operations and services through individual, private sector, nongovernmental, and public assistance programs that identify needs and define resources; provide housing and promote restoration; address long-term care and treatment of affected persons; implement additional measures for community restoration; incorporate mitigation measures and techniques, as flexible; evaluate the incident to identify lessons learned and develop initiatives to mitigate the effects of future incidents.

Red teaming—A group exercise to imagine all possible terrorist attack scenarios against the chemical infrastructure and their consequences.

Redundancy—An energy reliability strategy based on the notion that multiple systems provide needed backup if one system fails or cannot meet demand.

Reentrant corners—Internal or inside corners in the building envelope, usually at angles 90 degrees and greater.

Report printers—A separate, dedicated printer attached to the electronic security systems used for generating reports using information stored by the central computer.

Request-to-exit device—Passive infrared motion sensors or push buttons that are used to signal an electronic entry control system that egress is imminent or to unlock a door.

Resilience/resiliency—The ability to resist, absorb, recover from, or successfully adapt to adversity or a change in conditions. In the context of energy security, resilience is measured into terms of robustness, resourcefulness, and rapid recovery.

Resolution—The level to which video details can be determined in a CCTV scene.

Response—Activities that address the short-term, direct effects of an incident, including immediate actions to save lives, protect property, and meet basic humans needs. Response also includes the execution of emergency operations plans and incident mitigation activities designed to limit the loss of life, personal injury, property damage, and other unfavorable outcomes. As indicated by the situation, response activities include applying intelligence and other information to lessen the effects or consequences of an incident; increasing security operations; continuing investigations into the nature and source of the threat; ongoing surveillance and testing processes; immunizations, isolation, or quarantine; and specific law enforcement operations aimed at preempting, interdicting, or disrupting illegal activity, and apprehending actual perpetrators and bringing them to justice.

Response force—The people who respond to an act of aggression. Depending on the nature of the threat, the response force could consist of guards, special reaction teams, military or civilian police, an explosives ordinance disposal team, or a fire department.

Response time—The length of time from the instant an attack is detected to the instant a security force arrives on site.

Restricted area—Any area with access controls that is subject to these special restrictions or controls for security reasons.

Retentate—In ultrafiltration, the retentate is the solution that contains the particles that do not pass through the membrane filter. The retentate is also called the concentrate.

Retinal pattern—A biometric technology that is based on features of the human eye.

Ricin—A stable toxin easily made from the mash that remains after processed castor beans. At one time, it was used as an oral laxative, castor oil; castor oil causes diarrhea, nausea, vomiting, abdominal cramps, internal bleeding, liver and kidney failure, and circulatory failure. There is no antidote.

Risk—A measure of the potential for an unwanted outcome resulting from an incident, event, or occurrence, as determined by its likelihood and the associated consequences. Risk should be assessed from any scenario as a function of consequence, vulnerability, and threat, as defined below. It is important to think of risk as influenced by the nature and magnitude of a threat (T), the vulnerabilities to that threat (V), and the consequences that could result (C).

$$\text{Risk} = f\,(C,V,T)$$

- **Consequence (C) Analysis:** estimates the potential public health and economic impacts that a successful attack could cause
- **Vulnerability (V) Assessment**: identifies weaknesses in a asset design, implementation, or operations that can be explored by an adversary

- **Threat (T) Analysis:** estimates the likelihood that a particular target, or type of target, will be selected for attack, and is based on the intent and capacity of an adversary.

Rotating drum or rotating plate vehicle barrier—An active vehicle barrier used at vehicle entrances to controlled areas based on a drum or plate rotating into the path of the vehicle when signaled.

Sacrificial roof or wall—Roofs or walls that can be lost in a blast without damage to the primary asset.

Safe haven—Secure areas within the interior of the facility. A safe haven should be designed such that it requires more time to penetrate by aggressors than it takes for the response force to reach the protected area to rescue the occupants. It may be a haven from a physical attack or air-isolated haven from CBR contamination.

Salmonella—An infection caused by a gram-negative bacillus, a germ of the *Salmonella* genus. Infection with this bacteria may involve only intestinal tract or may be spread from the intestines to the bloodstream and then to other sites in the body. Symptoms of salmonella enteritis include diarrhea, nausea, fever, and abdominal cramps. Dehydration resulting from the diarrhea can cause death, and the disease could cause meningitis or septicemia. The incubation period is between eight and forty-eight hours, while the acute period of the illness can last for one to two weeks.

Sarin—A colorless, odorless gas. With a lethal dose of .5 mg (a pinprick-sized droplet), it is twenty-six times more deadly than cyanide gas. Because the vapor is heavier than air, it hovers close to the ground. Sarin degrades quickly in humid weather, but sarin's life expectancy increases as temperature gets higher, regardless of how humid it is.

Scramble keypad—A keypad on which the numbers change pattern with each use to enhance security by preventing eavesdropping observation of the entered numbers.

Secondary asset—An asset that supports a primary asset and whose compromise would indirectly affect the operation of the primary asset.

Secondary hazard—A threat whose potential would be realized as the result of a triggering event that of itself would constitute an emergency (e.g., dam failure might be a secondary hazard associated with earthquakes).

Sector—A logical collection of assets, systems, or networks that provide a common function to the economy, government, or society. The NIPP addresses 18 critical infrastructures sectors, identified by the criteria set forth in HSPD-7. This number has been condensed to 14 by Presidential Directive 21.

Sector/access mode—The state of an area monitored by an intrusion-detection system in regards to how alarm conditions are reported.

Security analysis—The method of studying the nature of and the relationship between assets, threats, and vulnerabilities.

Security console—Specialized furniture, racking, and related apparatus used to house the security equipment required in a control center.

Security engineering—The process of identifying practical, risk managed short- and long-term solutions to reduce and/or mitigate dynamic man-made hazards by integrating multiple factors, including construction, equipment, manpower, and procedures.

Semi-isolated fenced perimeters—Fence lines where approach areas are clear of obstruction for 60 to 100 feet outside of the fence and where the general public or other personnel seldom have reason to be in the area.

Sentinel laboratory—An LRN lab that reports unusual results that might indicate a possible outbreak, and refers specimens that may contain select biological agents in reference labs within the LRN.

Shielded wire—Wire with a conductive wrap used to mitigate electromagnetic emanations.

Shock wave—Highly compressed air traveling radially outward from the source at supersonic velocities.

Site characterization—The process of collecting information from an investigation site in order to support the evaluation of a drinking water contamination threat. Site characterization activities include the site investigation, field safety screening, rapid field testing of the water, and sample collection.

Situational awareness—An understanding of the current environment and the ability to accurately anticipate future problems in order to respond effectively.

Situational crime prevention—A crime-prevention strategy based on reducing the opportunities for crime by increasing the effort required to commit a crime, increasing the risks associated with committing the crime, and reducing the target appeal or vulnerability (whether property or person). This opportunity reduction is achieved by management and use policies such as procedures and training, as well as physical approaches such as alteration of the built environmental.

Sleeper cell—A small terrorist cell which keeps itself undetected until such time as they can "awaken" and cause havoc.

Smallpox—The first biological weapon, used during the eighteenth century, smallpox killed 300 million people in the nineteenth century. There is no specific treatment for smallpox disease, and the only prevention is vaccination. This currently poses a problem, since the vaccine was discontinued in 1970 and the WHO declared smallpox eradicated. Incubation is seven to seventeen days, during which the carrier is not contagious. Thirty percent of people exposed are infected, and it has a 30 percent mortality rate.

Smart card—A newer card technology that allows data to be written, stored, and read on a card typically used for identification and/or access.

Smart grid—The electric delivery network, from electrical generation to end-use customer, integrated with the latest advances in digital and information technology to improve customer, integrate with the latest advances in digital and information technology to improve electric-system reliability, security, and efficiency.

Sociopathic—Of, relating to, or characterized by asocial or antisocial behavior or a psychopathic (q.v.) personality, according to *Webster's*.

Specific threat—Known or postulated aggressor activity focused on targeting a particular asset.

Spore—An asexual, usually single-celled reproductive body of plants such as fungi, mosses or ferns; a microorganism, as a bacterium, in a resting or dormant state.

Standoff distance—A distance maintained between a building or portion thereof and the potential location for an explosive detonation or other threat.

Standoff weapons—Weapons such as antitank weapons and mortars that are launched from a distance at a target.

Stationary vehicle bomb—An explosive-laden car or truck stopped or parked near a building.

Strain-sensitive cable—Strain-sensitive cables are transducers that are uniformly sensitive along their entire length and generate an analog voltage when subjected to mechanical distortions or stress resulting from fence motion. They are typically attached to a chain-link fence about halfway between the bottom and top of the fence fabric with plastic ties.

Structural protective barriers—Human-made devices (e.g., fences, wall, floors, roofs, grills, bars, roadblocks, signs, other construction) used to restrict, channel, or impede access.

Superstructure—The supporting elements of a building above the foundation.

Supplies-bomb delivery—Bombs or incendiary devices concealed and delivered to supply or material handling points such as loading docks.

System—Any combination of facilities, equipment, personnel, procedures, and communications integrated for a specific purpose.

System events—Events that occur normally in the operation of a security management system. Examples include access control operations and changes of state in intrusion-detection sensors.

Tactics—The specific methods of achieving the aggressor's goals to injure personnel, destroy assets, or steal material or information.

Tamper switch—Intrusion-detection sensor that monitors an equipment enclosure for breach.

Tangle-foot wire—Barbed wire or tape suspended on short metal or wooden pickets outside a perimeter fence to create an obstacle to approach.

Target analysis—Estimates the likelihood that a particular target, or type of target, will be selected for attack, and is based on intent and capability of an adversary.

Taut wire sensor—An intrusion-detection sensor using a column of uniformly spaced horizontal wires, security anchored at each end and stretched taut. Each wire is attached to a sensor to indicated movement of the wire.

Technological hazards—Incidents that can arise from human activities such as manufacture, transportation, storage, and use of hazardous materials. For the sake of simplicity, technological emergencies are assumed to be accidental and their consequences are unintended.

Terrorism—Premeditated threat or act of violence against non-combatant persons, property, and environmental or economic targets to induce fear, intimate, coerce, or affect a government, the civilian population, or any segment thereof, in furtherance of political, social, ideological, or religious objectives.

Terrorist group—A group which practices or has significant elements which are involved in terrorism.

Thermally tempered glass (TTG)—Glass that is heat-treated to have a higher tensile strength and resistance to blast pressures, although with a greater susceptibility to airborne debris.

Threat—An indication that a harmful incident, such as contamination of the drinking water supply, may have occurred. The threat may be direct, such as a verbal or written threat, or circumstantial, such as a security breach or unusual water quality.

Threat analysis—A continual process of compiling and examining all available information concerning potential threats and human-caused hazards. A common method to evaluate terrorist groups is to review the factors of existence, capability, intentions, history, and targeting.

TNT equivalent weight—The weight of TNT (trinitrotoluene) that has an equivalent energetic output to that of a different weight of another explosive compound.

Toxic-free area—An area within a facility in which the air supply is free of toxic chemical or biological agents.

Toxic industrial chemical (TIC)—Manufactured for chemical industrial purposes. TICs are generally less toxic and less persistent than chemical warfare agents, and many are commonly stored or transported in bulk quantities.

Toxicity—A measure of the harmful effects produced by a given amount of a toxin on a living organism.

Toxin—A poisonous substance produced by living organisms capable of causing disease when introduced into the body tissues.

Transponder—A device on an airliner which sends out signal allowing air traffic controllers to track an airplane. Transponders were disabled in some of the planes hijacked on 9/11.

Transportation Security Administration (TSA)—A new agency created by the Patriot Act of 2001 for the purpose of overseeing technology and security in American airports.

Tularemia—An infectious disease caused by a hardy bacterium *Francisella tularensis,* found in animals, particularly especially rabbits, hares, and rodents. Symptoms depend upon how the person was exposed to tularemia but can include difficulty breathing, chest pain, bloody sputum, swollen and painful lymph glands, ulcers on the mouth or skin, swollen and painful eyes, and sore throat. Symptoms usually appear from three to five days after exposures but sometimes will take up to two weeks. Tularemia is not spread from person to person, so people who have it need not be isolated.

Twisted pair wire—Wire that uses pairs of wires twisted together to mitigate electromagnetic interference.

Two-person rule—A security strategy that requires two people to be present in or gain access to a secured area to prevent unobserved access by any individual.

Ultrafiltration—A filtration process for water that uses membranes to preferentially separate very small particles that are larger than the membrane's molecular weight cut-off, typically greater than 10,000 Daltons. (A Dalton is a unit of mass, defined as 1/12 the mass of a carbon-12 nucleus. It's also called the atomic mass unit, abbreviated as either "amu" or "u.")

Unobstructed space—Space around an inhabited building without obstruction large enough to conceal explosive devices 150 mm (6 inches) or great in height.

Unshielded wire—Wire that does not have a conductive wrap.

Value proposition—A statement that outlines the national and homeland security interest in protecting the nation's critical infrastructure and articulates the benefits gain by all critical infrastructure partners through the risk management framework and public-private partnership described in the NIPP.

Vault—A reinforced room for security items.

Vector—An organism which carries germs from one host to another.

Vertical rod—Typical door hardware often used with a crash bar to lock a door by inserting rods vertically from the door into the doorframe.

Vesicle—A blister filled with fluid.

Vibration sensor—An intrusion-detection sensor that changes state when vibration is present.

Video analytics—The use of computer software in surveillance to automatically identify things of interest such as behavior, objects, or attitude, without an operator having to view the video.

Video assessment and surveillance system(s) (VASS)—Digital monitoring and control systems to provide a rapid and cost-effective method for determining the source of the intrusion or other detection alarms.

Video intercom system—An intercom system that also incorporates as small CCTV system for verification.

Video motion detection—Motion detection technology that looks for changes in the pixels of a video image.

Visual surveillance—The aggressor uses ocular and photographic devices (such as binoculars and cameras with telephoto lenses) to monitor facility or installation operations or to see assets.

Voice recognition—A biometric technology that is based on nuances of the human voice.

Volumetric motion response—An interior intrusion-detection sensor that is designed to sense aggressor motion within a protected space.

Vulnerability—A physical feature or operational attribute that renders an entity open to exploitations or susceptible to a given hazard.

Waterborne contamination—CBR agent introduced into and fouling a water supply.

Weapons of mass destruction (WMD)—According to the National Defense Authorization Act: Any weapon or device that is intended, or has the capability, to cause death or serious bodily injury to a significant number of people through the release, dissemination, or impact of

- toxic or poisonous chemicals or their precursors
- a disease organism
- radiation or radioactivity

Weapons-grade material—Nuclear material considered most suitable for a nuclear weapon. It usually connotes uranium enriched to above 90 percent uranium-235 or plutonium with great than about 90 percent plutonium-239.

Xenophobia—Irrational fear of strangers or those who are different from oneself.

Zyklon b—A form of hydrogen cyanide. Symptoms of inhalation include increased respiratory rate, restlessness, headache, and giddiness followed later by convulsions, vomiting, respiratory failure and unconsciousness. Used in the Nazi gas chambers in WWII.

Index

www.ingramcontent.com/pod-product-compliance
Ingram Content Group UK Ltd.
Pitfield, Milton Keynes, MK11 3LW, UK
UKHW041142300926
15471UKWH00011B/102

9 781598 88994